AF711583

WEAVEXX

EN 29001/ISO 9001
ANSI/ASQC Q91
APPROVED BY BVQI (NA)

Frederic Emard
Design Engineer

1 Lee Blvd.
Warwick, Qc
J0A 1M0

Off: 1-800-374-5566 ext. 272
International: (819) 358-5566
Home: (819) 752-1042
Fax: (819) 358-5189

The Paper Machine Wet Press Manual

Third Edition, Revised

A Project of the Water Removal Committee of the Engineering Division

Edited by Peter Seifert

TAPPI
Atlanta, Georgia

TAPPI PRESS

The information and data contained in this document were prepared by a technical committee of the Association. The committee and the Association assume no liability or responsibility in connection with the use of such information or data, including but not limited to any liability or responsibility under patent, copyright, or trade secret laws. The user is responsible for determining that this document is the most recent edition published.

Within the context of this work the authors may use as examples specific manufacturers of equipment. This does not imply that these manufacturers are the only or best sources of the equipment or that TAPPI endorses them in any way.

The presentation of such material by TAPPI should not be construed as an endorsement of or suggestion for any agreed upon course of conduct or concerted action.

ISBN: 0-89852-056-8

3rd Edition, Revised

Copyright© 1991, 1979, 1972 by

TAPPI
Technology Park/Atlanta
P. O. Box 105113
Atlanta, GA 30348-5113 USA

All rights reserved

Permission of TAPPI is granted to photocopy items for internal or personal use of specific clients, for libraries or other users provided that the copying organization pay the base fee of $1.00 U.S. per copy, plus $.50 U.S. per page directly to the Copyright Clearance Center, 27 Congress Street, Salem, MA, 01970, U.S.A. (*ISBN*: 0-89852-056-8/1991) $1.00 + $.50 pp.

Library of Congress Cataloging-in-Publication Data

The Paper machine wet press manual / a project of the Water Removal Committee of the Engineering Division; edited by Peter Seifert. — 3rd ed., rev.

p. cm.

Includes bibliographical references.

ISBN 0-89852-056-8

1. Paper–Drying. 2. Papermaking machinery. I. Seifert, Peter, 1936– . II. Technical Association of the Pulp and Paper Industry. Water Removal Committee.

TS1118.D7P38 1991 91–27204

676'.232—dc20 CIP

Printed in the United States of America

B-042

Contributors

Authors of the Third Edition:

Aylor, Ronald L., JWI Group-Atlanta Felt
Berggren, Otto C., Huyck USA
Buck, Ronald J., Albany International
Busker, Leroy, Beloit Corporation
Curti, Jeffrey, Consolidated Papers, Inc.
DeCrosta, Edward Jr., dba Columbia Associates
Deuser, John, International Paper
Pikulik, Ivan I., Pulp and Paper Research Institute of Canada
Riggi, Jeff, Tampela-Sandy Hill, Inc.
Robinson, Kim Wm., James River Corporation
Seifert, Peter, Black Clawson Company
Smart, Alex, Black Clawson Company
Swenson, Roy S., International Paper Company
Vickery, Dennis, International Paper Company
Waller, Mike, Miami University
Walter, Sid H., Georgia-Pacific Corporation
Wise, Evan, Jefferson Smurfit/CCA

Other Current and Past Contributors:

Donna Adams, J. W. Anderson, David Antos, John Atkins, William Bliesner, Bernard Dahl, Fred Gedmer, Kenneth Jewett, Tom Kershaw, Robert A. Larson, Thomas Lauterbach, Al Mockaitis, Freeman Nichols, Bob Perrault, Richard Reese, Tom Sanders, Jack Smith, Frank Sutman, Larry Woodside, Joseph Zwinak

Preface

This book is intended to provide a brief introduction to the art and science of wet pressing in the paper industry, and also to aspects of day-to-day wet press operation in the paper mill environment.

The first and second editions served the 1970s and 1980s, the third edition of this book aims at the decade of the 1990s. Its contents have been fully updated, particularly to include new technology such as shoe presses, hot presses, and new felts. A chapter on personal safety was added, and also others, to describe new insights of press functioning and of the importance of pressing variables for water removal and generation of certain paper properties. Also, the structure of the text was simplified.

Proprietary names and trade names have been avoided, except in some cases where the reader is better served by using them. No endorsement is made of a certain product or supplier, and the reader is encouraged to look for additional designs or sources.

This book is a product of the Water Removal Committee of the TAPPI Engineering Division which hereby expresses sincere appreciation to the authors of the various chapters of this edition, and to the other current and past contributors.

Peter Seifert

CONTENTS

6. PRESS FABRICS

Illustrations

Tables

Chapter 1

Press Function

A. SHEET WATER REMOVAL ON A PAPER MACHINE

Papermaking consists of spreading a suspension of pulp evenly across a paper machine, followed by its dewatering in a manner that yields a product with desirable properties. Initially, in the forming zone, the water removal proceeds readily. However, as the solids content of the wet web increases, the additional water removal becomes gradually more difficult, time consuming, and expensive, in terms of both the capital and operating costs (Fig. 1.1). The web solids content can be increased in the former from less than 1% to about 13% to 23% by forces such as gravity, centrifugal force, hydraulic pressure between two forming fabrics, or vacuum created either by drainage elements of the former or by vacuum pumps.

The web solids content can be further increased by a mechanical compression of the web in the press section. Throughout its history, the industry has strived to maximize the water removal in the press section. On modern machines, 47% solids content can be achieved for newsprint and over 50% for a market pulp sheet. In some cases 55% dryness has been reached and we forecast that, by the end of the 1990s, even 60% values will be attainable.

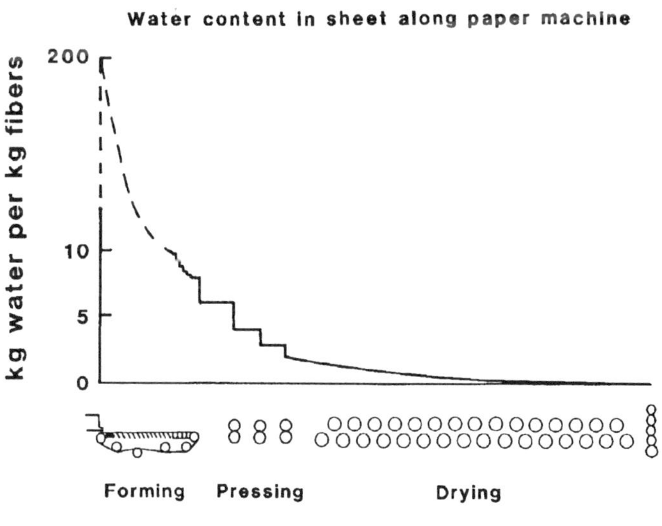

Fig. 1.1: Water removal along a paper machine

Water remaining in the pressed sheet must be eliminated by evaporation in a dryer section. Paper drying is a slow and expensive operation. Dryer sections of paper machines are large, costly, and require large amounts of energy. A wet sheet going through a long dryer section is at risk of damage or breaks.

Figure 1.2 shows the relative cost of water removal in the former, press section, and dryer section.

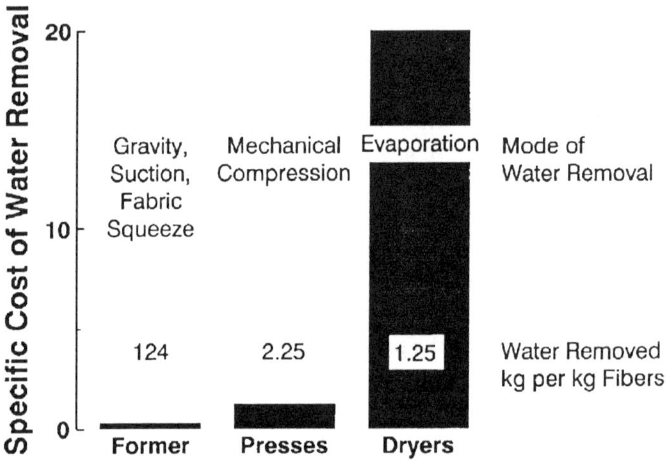

Fig. 1.2: Specific cost of water removal

B. WATER REMOVAL AND PAPER CONSOLIDATION IN A WET PRESS

The sheet leaving the former and entering the press section contains about 80% water. A portion of water carried by the web is located in the interstices between the fibers; the rest is contained in the lumen, pores, and swollen walls of cellulosic fibers. The bulk of this water is removed when the sheet is compressed in one or several consecutive press nips. **Figure 1.3** shows that a large portion of water carried by the sheet into the press section is removed in the first press. Although press nip loads normally increase from the first, to the second, to the third press, the amount of water removed decreases. As the sheets become dryer, removal of the remaining water becomes more difficult.

Pressing significantly augments the tensile strength of a wet sheet by raising the solids content. For example, a solids content increase from 17% to 40% results in about a threefold increase in the tensile strength of a newsprint sheet. Tensile strength is required to minimize sheet breaks, especially on fast machines producing low basis weight grades.

While pressing improves the economy of papermaking and the efficiency of the paper machine, perhaps its most important effect is improvement in the product quality. Pressing increases the sheet's smoothness, burst strength, tensile strength, and density, and reduces its thickness and tear strength. This factor cannot be overemphasized. For this reason an entire chapter has been added to this, the third edition of the manual.

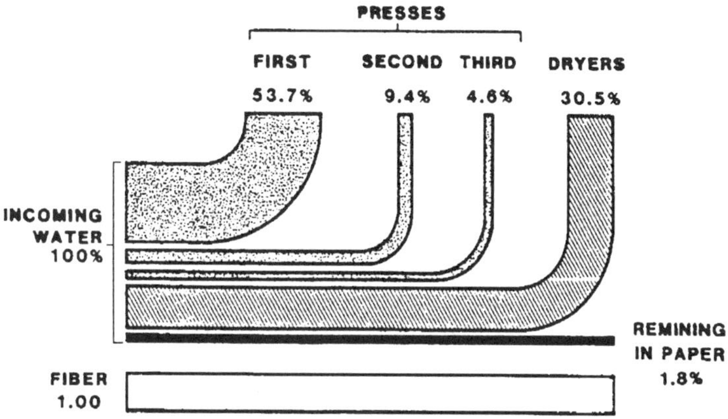

Fig. 1.3: Relative magnitudes of water streams (1)

C. MECHANISM OF WATER REMOVAL IN A PRESS NIP

In the press nip the wet web is compressed between two press rolls, of which one or both are felted. Water removed from the web in the press nip escapes either into the felt or through the felt into cavities provided in the press roll. On some slow machines it can escape through the incoming side of the nip.

Development of pressure during the passage of web and felt through the press nip has been qualitatively described by Wahlstrom (2), and later confirmed by experiments. The pressure distribution in the press nip is shown in Fig. 1.4. The pressure pulse initiates at the point where both the felt and the sheet contact the press roll. As the sheet and felt advance further into the nip, the pressure increases, reaches a maximum in the middle of the nip, and gradually drops to zero as the sheet separates from the felt on the outgoing nip side. The press roll pressure acting on the top side of the web is opposed by the mechanical pressure of the fibers and hydraulic pressure of the water contained in the web. In a single felted press, the hydraulic pressure in the sheet is greater on the side that faces the press roll than on the side that contacts the felt. This is because the felt offers relatively low resistance to a vertical water flow into the cavities provided in the press roll.

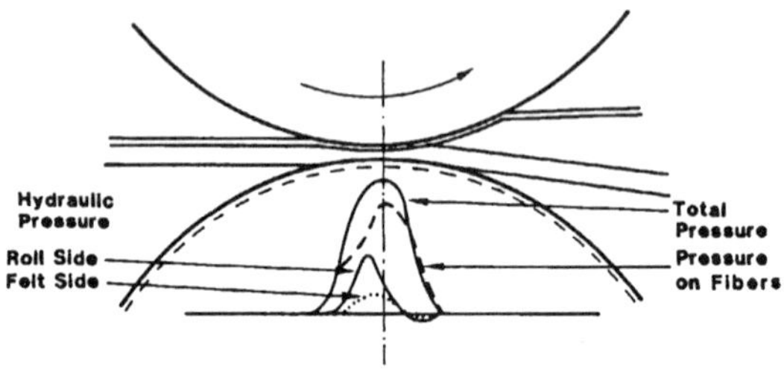

Fig. 1.4: Pressure distribution in a press nip

The force acting on the press roll is opposed by the mechanical pressure of fibers and hydraulic pressure of water between the fibers. Hydraulic pressure is greater on the roll side of the sheet than on its felt side.

The hydraulic pressure differential between the two sides of the web is the driving force for water removal. The extent of web dewatering is determined by the driving force, and the time available for water removal. The amount of water removed in the press nip is proportional to the area surrounded by the curves representing the hydraulic pressure on the two sheet sides in **Fig. 1.5**, or nip impulse. The mathematical expression of nip impulse is shown in equation (1).

$$\text{Nip Impulse} = \int_{t_b}^{t_e} (p_1 - p_2)\, dt \qquad (1)$$

t_b - time, pulse beginning
t_e - time, pulse end
p_1 and p_2 - pressure on the two sheet sides

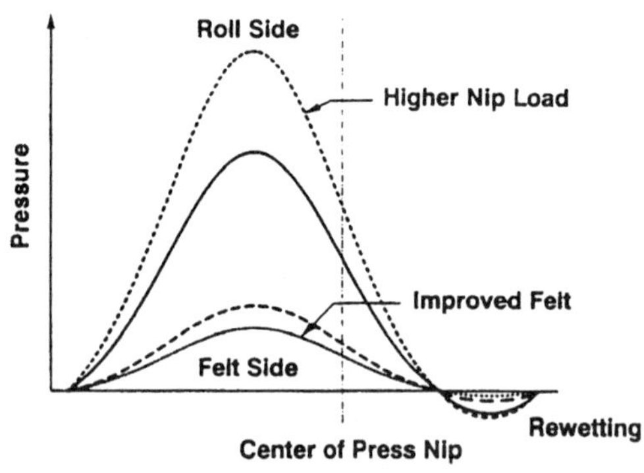

Fig. 1.5: Hydraulic pressure differential across the sheet

Water removal is proportional to the area between the two curves. Higher nip loads can improve the water removal.

Two simpler definitions of nip impulse are shown in equations (2) and (3).

Nip Impulse = (average nip pressure) x (nip residence time) (2)
Nip Impulse = nip load / machine speed (3)

Two ways of increasing water removal become evident on examination of **Fig. 1.5**. Application of a greater nip load increases the hydraulic pressure and the nip impulse, which drives the water removal. Installation of a more permeable felt or a more open press roll increases the water removal by reducing the hydraulic pressure on the bottom side of the sheet.

The nip residence time can be extended by the application of softer roll covers, more compressible felts, larger diameter press rolls, or the replacement of one press roll with a contoured press shoe. However, longer time alone has a relatively small effect, unless the total impulse is also increased.

D. ENTERING THE NIP, SHEET CRUSHING

Compression of the wet sheet on the entering side of the nip creates hydraulic pressure, and the pressure gradient across the sheet is the driving force for water removal. The hydraulic pressure reaches its highest value at the entering side of the nip where the sheet contains a large amount of water. By the time the sheet arrives in the nip's center, a large portion of its water has escaped into the felt and the hydraulic pressure is declining (see **Fig. 1.5**).

In addition to the hydraulic pressure gradient in the vertical direction - from the sheet, through the felt, to an open roll - hydraulic pressure gradient also develops in the horizontal direction. This horizontal pressure gradient causes a lateral water flow from the point of the highest hydraulic pressure forward towards the nip exit, where it can cause sheet rewetting, and backward towards the sheet entrance, where it can decrease the consistency of the incoming web and, by shear force, fluidize the fibrous mat. Sheet crushing results when a segment of the wet web is displaced towards the nip entrance.

The conditions which lead to sheet crushing include excessive hydraulic pressure in the nip and low web consistency. These occur most frequently in first presses, in plain presses where no vertical water escape route is available, or if water flow is obstructed by a heavy basis weight product, a plugged felt, or plugged cavities of the press roll.

A mild form of sheet crushing, so-called incipient crushing, leaves no visible marks on the sheet, but it can reduce significantly the sheet strength by rearrangement of its fibrous structure on a microscopic scale. More extensive crushing can cause visible sheet quality defects such as wrinkles and holes in selected areas of the sheet.

Crushing on a larger scale usually leads to a sheet break.

To eliminate crushing, steps have to be taken to minimize the hydraulic pressure in the first nip. For example, a reduction of the press nip load decreases the danger of crushing, but it also decreases the water removal. Other methods to minimize crushing include increasing the solids content of the incoming web, increasing felt permeability through improved cleaning, reducing felt moisture content, installing a new felt, or increasing the open surface in the press roll. Crushing potential increases with increasing moisture content, basis weight, press nip load, machine speed, and decreasing permeability of the sheet and felt.

E. LEAVING THE NIP, REWETTING

The exiting part of the nip extends from the middle of the nip to the point where the felt and the sheet leave the influence of the press rolls. In this region, the pressure applied to the sheet and the felt is gradually relieved, and their volumes expand. It is generally assumed that in the expanding part of the nip, a portion of water is transferred from the felt back into the paper through a process called internal rewetting. The driving force for the internal rewet is a negative pressure gradient, as shown in **Fig. 1.5**. The internal rewet is subject to all the variables that control water removal in the incoming part of the nip, such as the nature of the sheet and felt, and the pressure gradient. Widely varied estimates of the amount of internal rewet that occurs has been reported in the literature, but its real extent is not known.

On emerging from the nip, the web normally adheres to the plain roll. The geometry of the felt should be adjusted to assure an immediate separation of the felt from the web. If the sheet is allowed to remain with the felt after emerging from the nip, rewetting will continue. This external rewet is probably a result of capillary forces which draw water from larger pores of the felt to smaller pores of the paper. Water transfer from the felt to the sheet is opposed by the centrifugal force acting on the water on the perimeter of the rotating press roll. Therefore, the external rewetting is more extensive on slow machines in which the opposing centrifugal force is small. The amount of water transferred is independent of the basis weight. The low basis weight grades can therefore experience the largest decrease in percentage of the solids content. For example, external rewet can decrease the solids content of newsprint by as much as 3%.

In a regular operation, the web emerging from a press nip adheres to the roll. The web might adhere instead to the felt as the felt loses permeability through compaction, or as it becomes dirty. This "sheet stealing" can sometimes be prevented by wrapping the sheet by the felt over the solid press roll to enhance its contact with the smooth roll surface. On slow machines, such a geometry leads to an increased rewet.

References

1. I. I. Pikulik and I. T. Pye, Survey of press sections of Canadian pulp, paper and board machines. CPPA, Montreal, 1986.

2. B. Walstrom, Opportunities in pressing. Part I, Tappi J., 64(1):75 (1981) Part II, Tappi J., 64(2):57 (1981).

Chapter 2

Sheet Transfer To the Wet Press and To the Dryer Section

A. IMPORTANCE OF SHEET TRANSFER

The wet and very weak mat of paper must be transferred first from the forming part of the paper machine to the press section. After pressing, it passes on to the dryer section. Often, similar transfers are necessary within the press section itself. If the transfers don't work, no paper is made. If they fail frequently, excessive downtime occurs and the operating economy of papermaking suffers greatly. To gain an appreciation of the importance of these steps, the reader is encouraged to watch a start-up of a paper machine and to touch the wet paper to feel how weak it really is. **Figure** 2.1 describes how sheet strength develops as water is removed from the paper.

If the wet web is not continuously supported, the transfer or draw is open. If the paper is supported by a felt, the draw is closed.

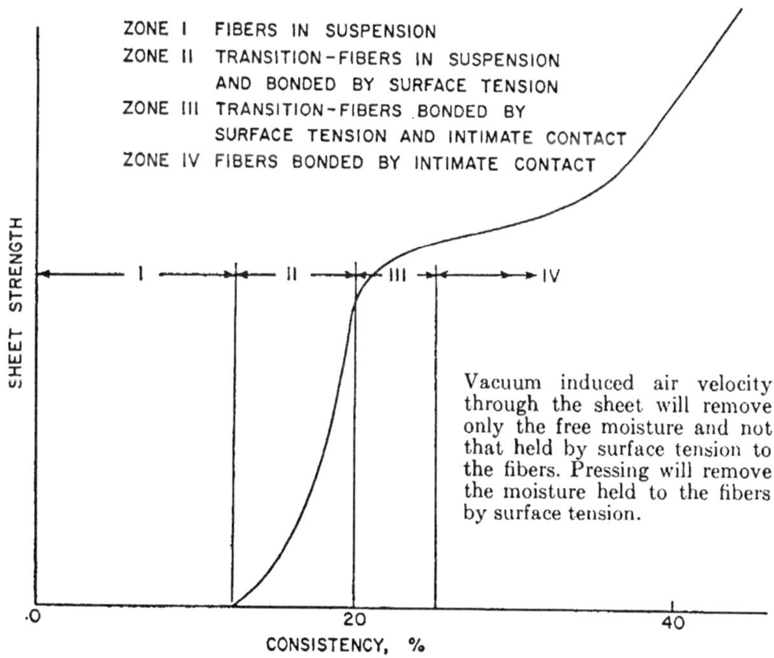

Fig. 2.1: Variation of sheet strength with consistency

B. TRANSFER FROM FORMER TO WET PRESS

Open Draw

Open draws between fourdriniers and press sections, as shown in **Fig. 2.2**, are suitable only for slower speeds, up to 2000 ft/min, depending on basis weight and sheet wet strength. However, in some isolated instances, even higher speeds have been achieved. Some issues and limitations of operating machines with open draws are described below and are also indicated in **Fig. 2.2**.

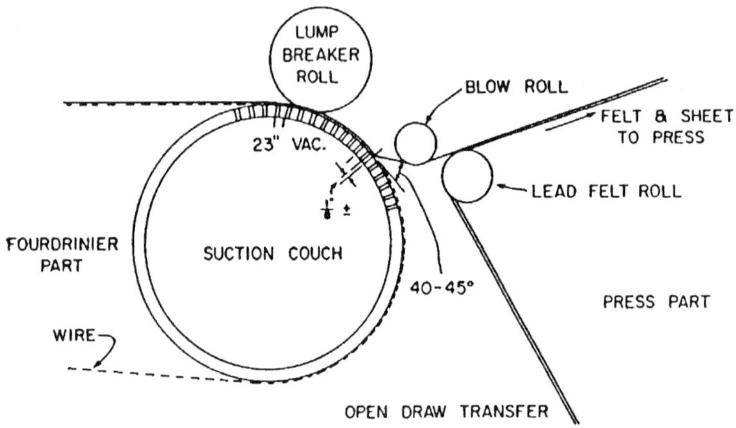

1. Typically, to attain a dyness of 20% couch vacuum must be about 23" hg. Lump breaker roll makes this possible by increasing dryness about 2% to 3%. Generally, below 19% dryness sheet does not have strength to be pulled off couch.
2. Sheet is pulled off couch at point following suction box seal in zone of diminishing vacuum. If sheet follows couch it will be rewetted by water thrown out from suction holes and sheet will break.
3. If angle of take-off is decreased, the sheet will follow couch and be rewetted, if increased it will draw off in high vacuum area and break.
4. Generally, safe operating speed is limited to 1,500 to 2,000 fpm depending on sheet wet strength.

Fig. 2.2: Open draw transfer

At machine speeds above 1000 ft/min the water removed from the sheet is drawn as a mist together with air into the suction holes of the couch and not all the way into the suction box. As the holes pass from the suction box area to the area of the sealing strips and beyond, the air velocity diminishes to zero. Water held in the holes is then released. The sheet must be pulled off the wire at a point of diminishing couch vacuum to avoid rewetting by the water thrown out of the holes. The sheet must have sufficient strength, which normally requires a consistency above 19%, so that it can be pulled away from the wire. If the sheet is too wet and too weak, it will break.

The angle at which the sheet is pulled off the couch with respect to the tangent at the point of takeoff should be around 40° to 45°. If it is too great, a slight change in draw will cause the sheet to be pulled off at a point of higher vacuum and the sheet will break. Examination of the break will show the couch suction hole pattern with fibers pulled out of the sheet in the areas of the holes. If the takeoff angle is too acute, the sheet will tend to follow the couch even with draw increased to the breaking point. This will allow the sheet to be rewet by the water thrown out of the couch holes. Sheet breaks will also occur if the takeoff angle is correct but the draws are soft because of a soft stock condition; in this case sheet will follow the couch, rewet, and then break.

A blow roll (see Fig. 2.2) can stabilize the point of sheet takeoff, thereby providing better operation. The takeoff angle is controlled by the blow roll position with respect to point of takeoff. A lump breaker roll riding on the sheet over the couch suction box area will compact the sheet and reduce its permeability. The higher vacuum thus achieved in the couch roll, together with the lump breaker's pressing action, increases sheet consistency and aids in keeping it above the critical value of 19%. However, a lump breaker roll can increase wire marking.

Poor Man's Pickup

This arrangement employs a wire turning roll. After the couch, the sheet follows the wire. The couch suction box is positioned so that water released from the couch is not thrown toward the leaving wire (Fig. 2.3).

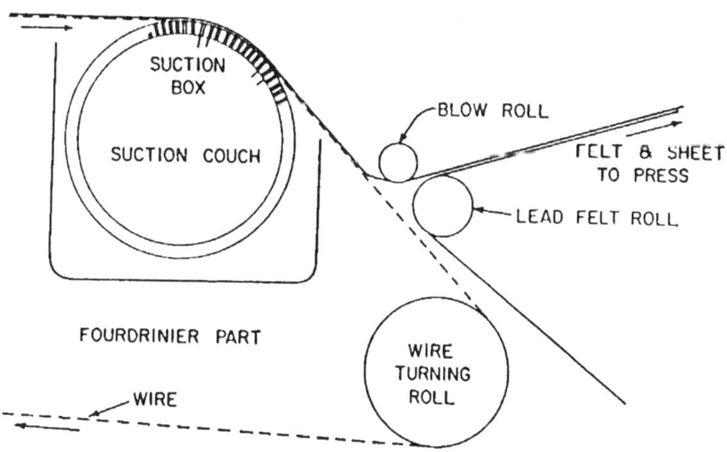

Figure 2.3 Poor Man's Pickup

The first press bottom felt lead-in roll is positioned close to the wire about midway between the couch and wire turning roll. The sheet is pulled off the wire and carried on top of the felt to the press. Sometimes, a blow roll is used to stabilize sheet takeoff from the wire.

The only restraining force that must be overcome by the wet paper is the adhesion of the water film on the wire. As a result, this takeoff is suitable for slightly higher speeds than the open draw. It has been used successfully on kraft paper and board machines at speeds in excess of 2000 ft/min.

Suction Pickup

The suction pickup is a closed draw arrangement in which the wet paper remains on the wire until after the couch. The top felt of the first press virtually touches the forming fabric or wire as it runs around the suction pickup roll, which transfers the sheet from the wire to the bottom side of the felt. The couch suction box is positioned so that the water thrown from the couch suction holes does not reenter the wire, so that rewetting of the sheet is avoided. The wet paper is supported by the felt, so much lower consistencies may be tolerated, down to 10%, on occasion. There is no speed limitation. The sheet, now attached to the underside of the pickup felt, is transferred to the first press, which may be single- or double-felted.

Figure 2.4 shows a suction pickup, **Fig. 2.5** a combined suction pickup and first press. Other arrangements are discussed in Chapter 4 of this manual.

Yankee Pickup

The Yankee pickup is a closed draw transfer, designed for paper machines that produce very light tissue at very high speed. It consists of a grooved couch roll and a felted pickup roll directly in contact with the couch (**Fig. 2.6**). The pickup felt is sufficiently wet so that the sheet adheres to it. Sometimes, the felt pickup roll is installed slightly ahead of the couch roll and set down into the wire run. This version is called free wire pickup. Sheet consistency leaving the wire is between 10% and 15%.

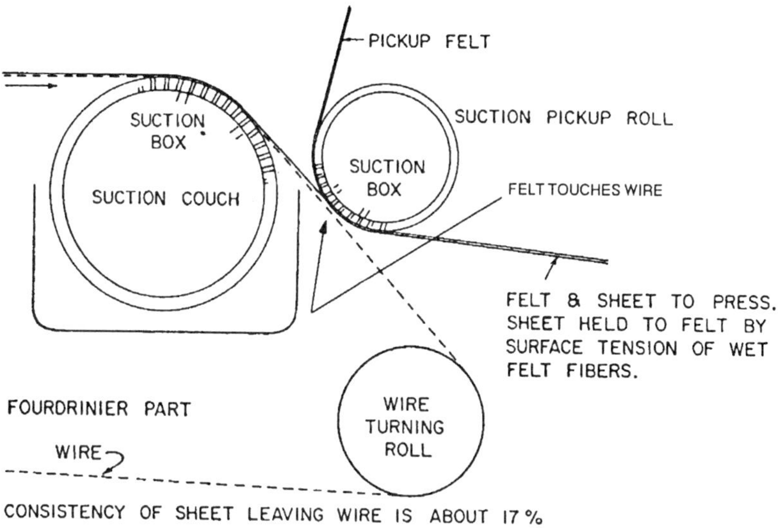

Fig. 2.4: Suction pickup

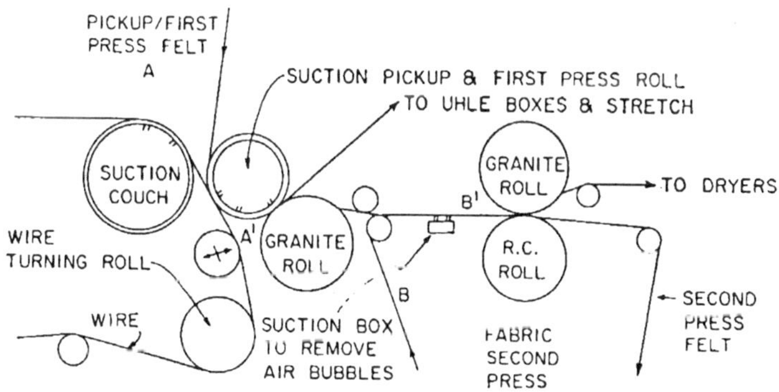

1. Relatively dry pickup/first press felt permits high nip loading and high moisture removal.
2. Second press can have grooved roll, blind drilled, or suction roll arrangement.
3. With arrangement shown a sheet dryness of 36/38% can be had leaving first press and 42/44% leaving second press without sheet damage.
4. Pick-up is also commonly achieved by moving the wire turning roll or the suction pick-up assembly.

Fig. 2.5: Combined suction pickup and first press

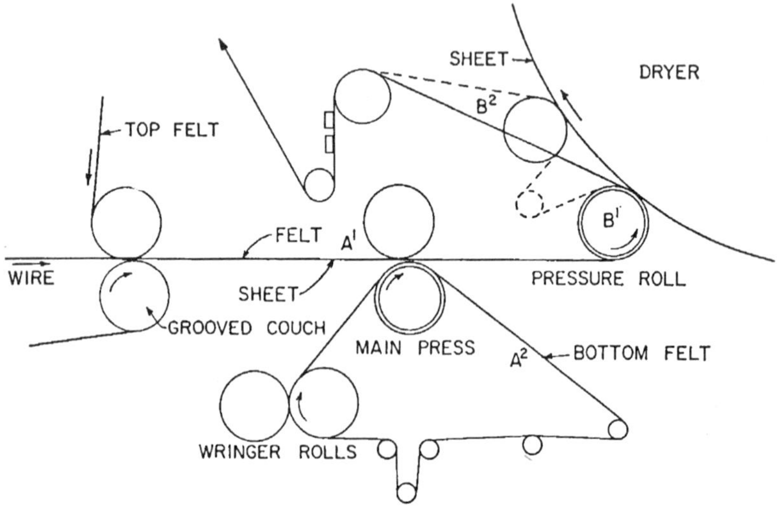

Configuration can be with one or two pressure rolls or the yankee dryer.
Note two suction boxes in tip felt run.

Fig. 2.6: Yankee pickup

The free wire pickup may also feature a stationary foil in place of the pickup roll guiding the top felt into contact with the wet paper (Fig. 2.7). Such an arrangement is in common use today because of desirable characteristics, such as:

- Quicker break-in time on new pickup felt
- More uniform pickup
- Minimum maintenance because there are no rotating members
- Greater unit pressure at the pickup point due to a small pickup radius
- Cleaner separation of the sheet from the wire due to the vacuum action of the foil

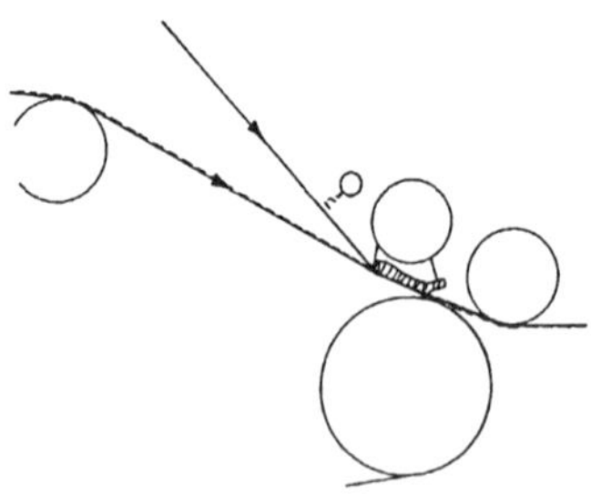

Fig. 2.7: Tissue foil pickup

In any of these plain pickup arrangements i.e., without suction rolls, the felt's ability to effect a clean pickup and retain the sheet on the underside of the felt is directly related to the felt's moisture content and its distribution in the felt. Therefore, it has always been necessary to use a lot of water on the felts and remove it again at the first suction pressure roll.

If it is desired to send less water to the first press, a suction pickup as in **Fig. 2.8** could be used.

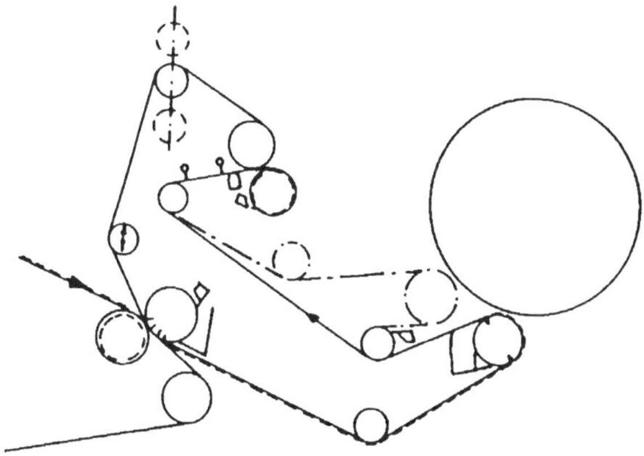

The wringer press shown here is not commonly used any more.

Fig. 2.8: Tissue suction pickup.

C. TRANSFER FROM THE PRESS SECTION TO THE DRYERS

In many paper machines today, draws between the forming zone and the press and within the press section have been closed, while the transfer from the press section to the dryer section is usually open. This arrangement permits start-ups of the forming section and press section, while the full width sheet after the press can be recycled until the wet end is stabilized. Only then is the paper threaded through the dryer section.

However, within the press section after single felted presses, there is always an open draw as shown twice in **Fig. 2.5**. The paper is unsupported and must be peeled away from the surface of the rolls. Granite is a most suitable roll surface material and is widely used because it has good release characteristics and requires only minimum peeling forces. Granite substitutes are being developed and are increasingly accepted. In the 1990s, and beyond, a trend is expected in which granite rolls will be replaced with metal rolls covered with easy-release, wear- resistant materials.

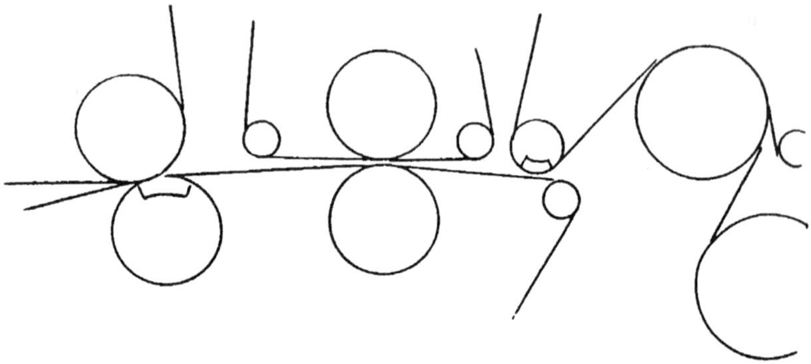

Press section without open draws, first top felt coming from suction pick-up. Last press is double felted, allowing suction roll transfer of the sheet to the dryer fabric.

Fig. 2.9: Closed draw to dryer section.

If the last press is double felted as shown in **Fig. 2.9**, the paper can be pulled easily from the felt to the dryer section. Adhesion between the felt and the paper is not very strong.

If the last press is single felted, as in **Fig. 2.10**, the paper tends to follow the plain roll and must be peeled off from its surface. The paper should be dry enough so that it has sufficient tensile strength to permit the continuous peeling. Factors like surface roughness, sticking tendency of the paper, and release angle have been the subject of many practical and scientific studies. The above remarks about granite rolls and new granite substitutes apply here, too.

Many attempts have been made to close the draw by means of a suction pickup arrangement directly on the plain roll. So far, with some exceptions, such attempts have not been successful. Closing draws is a significant benefit for paper machine economy. Developments and actual commercial experience with them should be observed eagerly and, if promising, should be adopted.

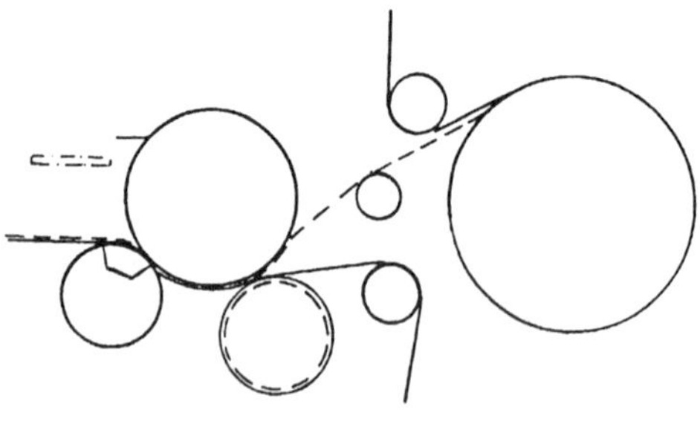

Fig. 2.10: Open draw to dryer section.

Chapter 3

Press Types

Eight general types of wet presses are used on paper machines today. The limitations, advantages, and disadvantages of each are described in this chapter.

A. PLAIN PRESS

Many years ago, wet presses on all paper machines were of the plain press type. Its use today is generally limited to slower machines (speeds of 1000 ft/min and below) in the second and third press positions. **Figure 3.1** illustrates a plain press.

The plain press operation is similar to that of the wringer on an old-fashioned washing machine. The felt and sheet pass through the nip between two solid press rolls. The water squeezed out at the nip flows backward, opposite to sheet travel and the rotation of the rolls, down the surface of the bottom press roll, and off into a saveall pan. To minimize saturating the felt ahead of the nip and thereby rewetting the sheet, the top roll is couched ahead of the bottom roll so that the felt and sheet wrap the top roll. This creates an open space under the felt so water can flow back unrestricted on the surface of the bottom roll. The wrap of the sheet and felt on the top roll also helps to force out any air between the sheet and felt, preventing blowing and press wrinkles.

At higher machine speeds, the difference in velocity between the water flowing backward out of the nip and the sheet traveling through the nip becomes great enough to make the force of the water against the sheet fibers exceed the strength of the fiber network. This causes the sheet to crush. In some instances, the sheet and felt wrap of the top roll is inadequate to remove all the air trapped between the sheet and felt, and the sheet will contain blow wrinkles. To ensure more complete air removal, a slotted suction box can be positioned ahead of the press. To prevent blowing, the sheet can be fed over a paper roll ahead of the press so that it contacts the top roll before the felt.

As with all press types, the sheet and felt are separated at the nip exit to prevent sheet rewet, the reabsorption of water from the felt. The felt should wrap the bottom roll and the sheet should be allowed to follow the top roll. In some instances the sheet is permitted to fall back onto the felt, but this should be avoided if possible. It is better to position the succeeding press above the felt of the exiting press so the succeeding press felt can then convey the sheet out of contact with the exiting press felt. The succeeding felt supports the tender sheet, providing safer operation. A paper roll is often positioned after the nip to stabilize the take-off of the sheet from the top roll, which also contributes to safer operation.

B. SUCTION PRESS

The first improvement in press design was the development of the suction press (**Fig. 3.2**). Prior to the use of even more modern presses such as the grooved roll, fabric, shrink sleeve, and impulse presses, this was the only type employed on high-speed paper machines.

The suction press roll is composed of a bronze or stainless steel shell covered with either rubber or a synthetic material such as polyurethane. Both the shell and cover have approximately 1/8 in. diameter holes drilled on about 5/16 in. centers over the entire roll surface. A nonrotating suction box fitted to the inside surface of the shell extends across the face. Air or spring-loaded seals are positioned between the inside shell surface and the box. Liquid-ring type vacuum pumps or centrifugal exhausters, usually located in the machine room basement, provide the vacuum.

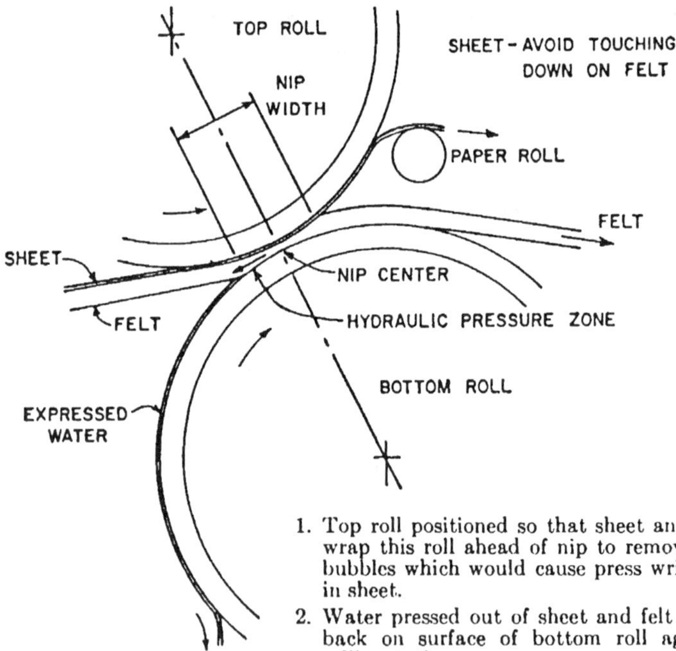

1. Top roll positioned so that sheet and felt wrap this roll ahead of nip to remove air bubbles which would cause press wrinkles in sheet.
2. Water pressed out of sheet and felt flows back on surface of bottom roll against roll's rotation.
3. Plain press limited to 1000 ft/min machine speed.

Fig. 3.1: Plain press

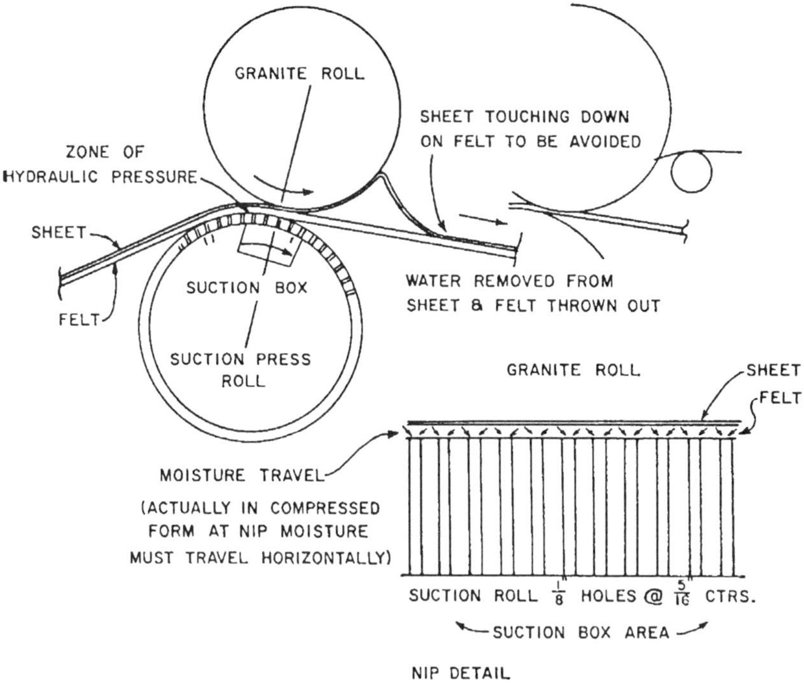

1. Felt and sheet wrap suction zone to remove air bubbles and to seal suction zone.
2. Water from sheet and felt drawn into holes of bottom roll by vacuum and hydraulic pressure.
3. Water must travel both vertically and horizontally through felt to holes in bottom roll.
4. Sheet is shown contacting felt after leaving press. This should be avoided because of rewetting of sheet.

Fig. 3.2: Suction press

The suction box is positioned in the roll so that the inside edge of the trailing seal strip is slightly beyond the center of the nip formed by the suction roll and the solid press roll, yet still within the nip area. The suction area is located just ahead of the nip. Felt and sheet travel is arranged to wrap the suction area so that their surfaces at the in-going nip are under vacuum. This provides a seal against air leakage and removes any air trapped between the felt and sheet, preventing blowing and felt wrinkles. As the felt and sheet pass through the nip between the suction and plain rolls, water is squeezed out at the in-going nip. Capillary action of the felt fibers and the hydraulic pressure developed by the compression of the felt and sheet in the nip cause water to flow in and through the felt and into the holes of the suction roll. In some instances the pressure gradient developed causes the water to flow with such velocity that a migration of fines occurs in the areas of the sheet over the suction holes. This condition, referred to as shadow marking, usually can be corrected by using a different construction felt, lower nip pressure, or a lower vacuum.

At machine speeds below approximately 1000 ft/min, water is pulled through the suction holes and into the vacuum box. At higher speeds, centrifugal force allows the water into only the holes.

To reduce rewetting of the felt at the outgoing nip, the vacuum box on some press installations is positioned to allow a slight air bleed between the nip and the inside edge of the trailing strip of the box. The felt leaving the nip is directed so that it does not wrap the suction roll, but is tangent to the press rolls at the line of the nip. When the sheet and felt leaving the nip are separated, the sheet is permitted to adhere to the solid roll leaving the nip. The sheet is then pulled off the solid roll, often stabilized by contact with a paper roll, and then drawn into the succeeding press nip.

The distance that the water must travel to be removed at the suction roll holes can create considerable hydraulic pressure in the nip. This can result in a condition where the volume of water leaving the holes of the suction roll is less than that which is entering the press and must be accommodated by the void volume within the nip. This causes water in the sheet to flow backwards out of nip, leading to crushing. The degree of crushing depends on the actual volume and flow velocity of water, felt construction and condition, nip pressure, and vacuum in the suction roll. If the inside edge of the trailing seal strip of the internal vacuum box is not positioned far enough out of the exiting side of the nip center, crushing will be extensive. Obviously, this is to be avoided.

Two actions occur as the suction holes leave the nip. First, air rushes into the exposed holes that are open to the vacuum chamber or sealed by the trailing strip which prevents water from leaving. Second, as the holes leave the strip and the vacuum is broken, the water is thrown out of the holes by centrifugal force. The lip of the saveall pan should be positioned close to the felt so that the water thrown out of the roll flows into the pan rather than rewetting the felt. The saveall pan lip usually is fitted with an adjustable strip.

The high cost of electric power following the energy crisis of the 1970s caused paper manufacturers to look for ways to reduce their energy expenditures. A resulting innovation was the use of so-called blind drilled press rolls, which are essentially suction rolls without the suction. The roll cover is drilled just like a suction roll but the shell is not. A vacuum source is not needed. Their limitation is less water-handling capacity than conventional suction rolls, which has restricted their use to positions that do not require high hydraulic loads. Blind-drilled press rolls are useful in double-felted presses and second presses on lighter-weight grades.

C. GROOVED ROLL PRESS

Nearly all paper machines built in the past 20 years have been equipped with grooved roll presses (**Fig. 3.3**).

This roll has circumferential spiral grooves (usually 0.020 in. wide by 0.110 — 0.120 in. deep at 0.125 in. spacing) machined into the cover. The water squeezed out in the nip by hydraulic pressure is forced into the grooves. The centrifugal force at the outgoing side of the nip then slings out the water in the grooves.

A deposit of only a few thousandths of an inch of fines, pitch, or aluminates in the grooves will greatly reduced its water-removal capacity. This condition will often result in crushing, so these rolls usually are equipped with purge showers to keep the grooves free of deposits.

The solid roll often is positioned as in a plain press, slightly before the grooved roll so that the sheet and felt contact the solid roll first to aid in preventing blowing and consequent blow wrinkles. To further prevent blowing, the sheet should contact the solid roll before the felt ahead of the nip.

The distance water must travel horizontally in the felt to the grooves is much less than the distance it must travel to the holes in a suction roll. Therefore, hydraulic pressure will not be as great and there will be less tendency for sheet crushing and shadow marking. Although grooved roll marking can occur, careful attention to felt conditioning to retain its water-handling capacity will largely prevent its occurrence. Advanced-design felts have multilayer, noncompressible base fabrics and high water handling ability and have greatly diminished the incidence of grooved-roll marking.

The grooved roll is an efficient press that accommodates high nip loading with a high rate of water removal.

Although the grooved roll press has less water-handling capacity than a suction press, its simplicity and ease of operation make it economically attractive.

A grooved roll requires a much smaller capital investment than a suction press roll. A vacuum pump, piping, and special framing are not needed; the expensive suction roll and its high maintenance cost are eliminated; and the reduced requirement for vacuum also lowers operating costs.

A grooved roll can be equipped with a variable crown mechanism which allows the press load to vary as needed, however, this adds to the capital and operating costs.

D. FABRIC PRESS

The fabric press has been popular in Europe (**Fig. 3.4**).

Except for its clothing, the fabric press is identical to a plain press: Both press rolls are solid. A multilayer weaved fabric belt passes through the press nip sandwiched between the rubber-covered roll and the felt. This compression-resistant inner fabric has a relatively high capacity to receive and remove water squeezed out at the nip. Hydraulic pressure developed at the in-going nip is relatively low. The fabric press will tolerate a fairly wet felt, high nip pressures, and low-bulk felt without crushing. Water is removed by passing the fabric over a vacuum box on its return run to the nip.

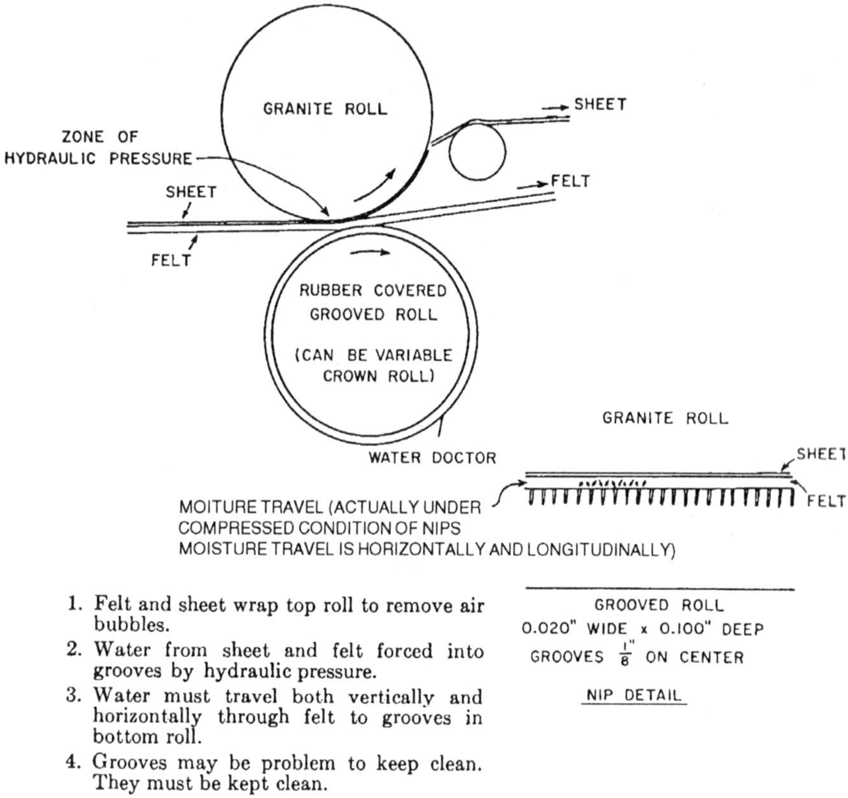

Fig. 3.3: Grooved roll press

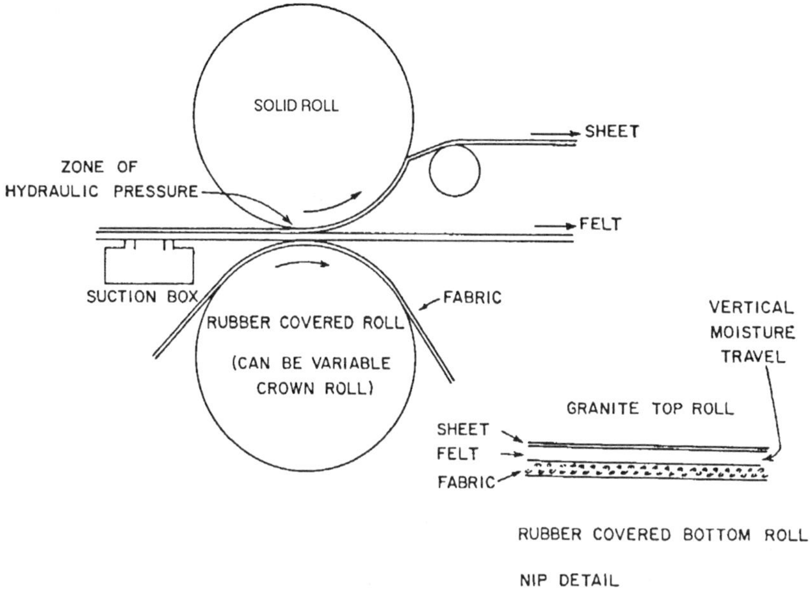

1. Suction box ahead of nip removes air bubbles. An alternate is having sheet and felt wrap top roll.
2. Hydraulic pressure forces water pressed out of sheet and felt into fabric. Little hydraulic pressure developed.
3. Fabric is multiple weave with the interstices of weave capable of holding considerable water.
4. Fabric is passed over suction box to remove water.

Fig. 3.4: Fabric press
Fabric presses are less frequently used.

Concerns about excessive felt wear, in addition to the added cost of the fabric and its operation, have largely prevented the use of the fabric press by U.S. paper manufacturers.

E. SHRINK SLEEVE PRESS

The shrink sleeve press is a simplified version of the fabric press (**Fig. 3.5**). A fabric jacket or sleeve is shrunk over the rubber-covered press roll. The press may or may not be identical to a plain press. About 40% of the installations were specifically undertaken to reduce shadow marking on existing suction presses, with no major press modifications.

A problem with this type of press is the fabric's tendency to creep on a roll with a large crown or high nip pressures. This can cause a badly distorted weave pattern and fabric failure. Hard press rolls, at 20 P&J Plastomer of lower, are recommended to overcome this problem.

F. HIGH IMPULSE PRESSES

Another result of the effort to stem rapidly escalating energy costs during the 1970s energy crisis was the recognition that the amount of water removed from the sheet in a press nip is a function of the "press impulse". This is defined as the product of the average pressure and dwell time in the nip, or the area under the pressure vs. time curve (See **Fig. 1.4**, p. 4, or **Fig. 10.1**, p. 98,) from nip entrance to nip exit. Much higher sheet dryness than that resulting from conventional pressing practice could be achieved by increasing the nip residence by several times. This finding led to the development of presses with exceptionally wide nips and high nip loads. These presses have the generic name high impulse presses.

Shoe presses and jumbo roll presses are the two basic configurations of high impulse presses.

The shoe-type impulse press shown in schematic in **Fig. 3.6** is a single grooved press roll opposed by a stationary loaded shoe. The press is double felted, with one felt against the top roll and the other passing through the nip surrounding the shoe and its supporting beam. The paper web passes through the nip between the two felts. Between the shoe and the felt is an impervious elastomer belt. This is lubricated by oil to create a slipper bearing between the shoe and the inside of the belt; it also permits force to be applied to the felt and paper sandwich and the opposing roll. The shoe is loaded against the belt with a hydraulic piston underneath extending across the full width of the machine. The load averages approximately 600 psi (pounds per square inch) along the full length of the shoe, which is equivalent to approximately 6000 pli (pounds per lineal inch) by the normal method of stipulating press loadings. Although the 600 psi nip pressure is not excessive with respect to the sheet and felt and is less than the peak pressure reached in many conventional roll presses, it is sustained five to eight times longer in the nip. The extremely large forces resulting from the high loading necessitated a change in press framing to support them.

The sheet that leaves the press is significantly dryer and stronger than one leaving a conventional press. The following benefits have been claimed:

- 45% dryness on virgin linerboard
- improved uniformity of dryness as a result of longer time in the nip
- increased physical properties such as Mullen burst, ring crush, and tensile strength with a slight reduction in caliper and porosity.

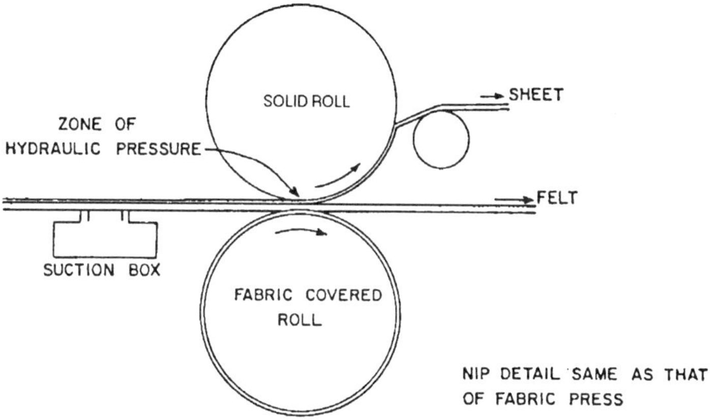

1. Shrink fabric press is simplified fabric press. Cheaper to install and requires less space, shorter fabric life, greater felt wear, longer changing time. Dewatering capacity same.

2. Existing presses are easily converted to shrink fabric presses. About 40% of conversions have been on suction presses resulting in less shadow marking and crushing.

Fig. 3.5: Shrink fabric sleeve press
Use is very rare.

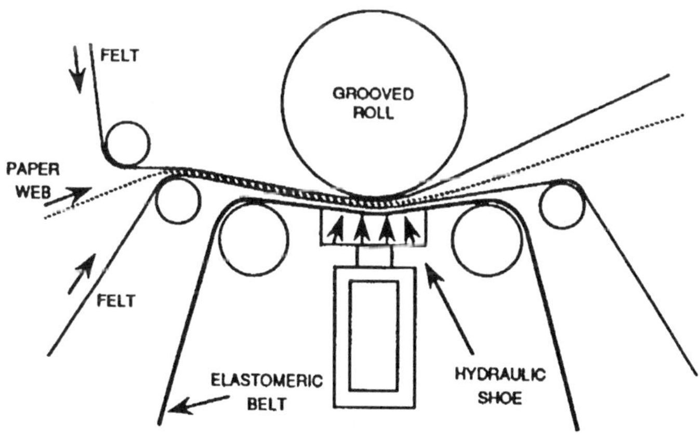

Fig. 3.6: Shoe-type high impulse press, with belt
The nip is formed between shoe and top roll which may be grooved.
The belt glides over the shoe, libricated by oil. While average static pressure in the nip is not very high, together with the length of the nip, very high linear loads are possible.

This type of press has several possibilities for papermakers:

- use of lower strength pulps for a given sheet strength
- reduced refining with a given pulp
- increased production off dryer limited machines
- reduction in dryer section length for a given tonnage.

On an existing machine, a 20—25% reduction in energy for drying paper can be expected.

The other shoe-type impulse press is shown in **Fig. 3.7**. The load is also applied with a hydraulic piston, but the shoe and loading piston are contained within a roll-like device with a flexible shell that conforms to the shape of the shoe and opposing roll. This configuration is less likely to contaminate the process because the lubricating oil for the shoe and inside surface of the roll's flexible sheet is contained within the closure of the roll. Very high loadings of several thousand pli can be applied; the very wide nip keeping the pressure well within the range of conventional presses. The combination of wide nip and high pressure results in sheet dryness similar to that achieved on the other shoe-type press.

The jumbo roll press is the second type of high impulse press. This press utilizes very large diameter rolls, one of which has a relatively soft cover to obtain a wide nip. The nip widths are not as wide as those of the shoe-type presses, and the loads are not as high. However, the resulting product of the nip width and average pressure is beyond conventional practice, which qualifies it as an impulse press. The results achieved tend to be intermediate between conventional presses and shoe-type presses. These presses are an attractive option to paper manufcaturers because of their simplicity and lower capital cost compared to the shoe-type press.

G. HOT PRESSES

The direct action hot press also resulted from the efforts to reduce energy costs for paper manufacturing. It has been long known that the rate of fluid flow through a porous medium is inversely proportional to the viscosity of the fluid, and that the viscosity of water decreases rapidly as its temperature increases. Many mills have added steam showers in the press section to raise the temperature of the sheet and the water it contains to achieve increased water removal.

The direct action hot press, shown in **Fig. 3.8**, was designed to increase the sheet temperature directly in the press nip. A large-diameter steel or cast iron center roll coated with a special hardened alloy is internally heated with steam at pressures up to 50 psi. The sheet contacts this roll directly, thereby being heated at the same time that it is placed under pressure. Excellent heat transfer is achieved resulting in excellent water flow and drainage through the porous structure of the sheet and felt.

Most applications are configured with two or more press nips providing maximum utilization of the steam-heated center roll. The press generally is designed as a nondraw press. Nip loads range from 150 to 1000 pli. These moderate loads and temperatures do not demand special felts or roll cover properties, but they extend the technology of conventional high impulse and wide-nip presses to achieve levels of press dewatering never before possible. Sheet drynesses from 50% to 57% can be obtained.

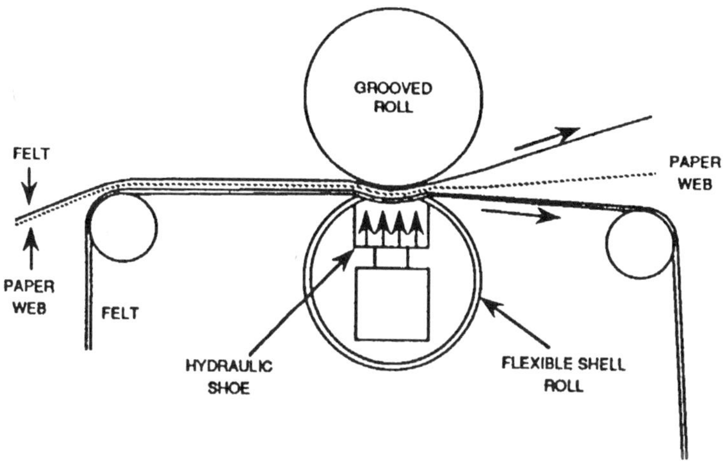

Fig. 3.7: Shoe-type high impulse press, with flexible shell

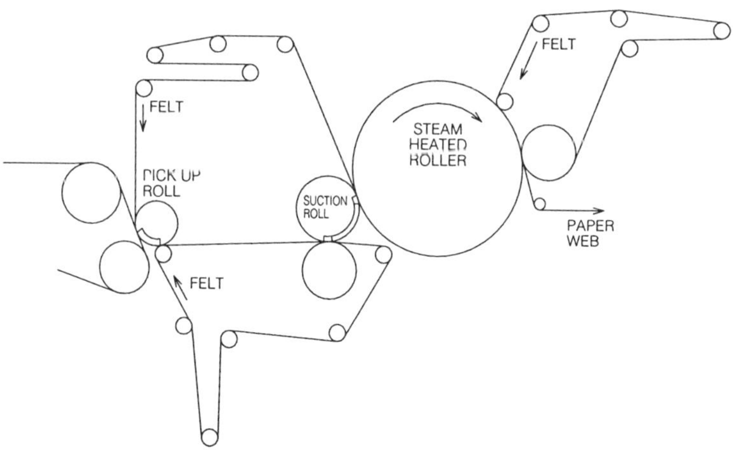

Fig. 3.8: Direct action hot press

Improved physical properties of the paper and board have been reported. These include:

- 10 – 15% increases in tensile and burst strengths
- improved stiffness with no change in porosity or permeability
- improvements in tear strength and linting tendency.

These improvements provide options for increasing productivity and reducing operating costs.

Chapter 4

Press Arrangements

While the previous chapter describes the various press types, this chapter concentrates on press arrangements which are combinations of the various types. Generally, arrangements are classified according to how many rolls are used, i.e., two-roll or three-roll arrangements. All have the primary function of water removal, and also secondary ones such as the transfer of the web from one felt to another, smoothing the web, surface, or otherwise affecting paper properties.

In the last 20 years (1970 to 1990), press arrangements have been revolutionized and have contributed greatly to savings in dryer energy, improved press runnability, and improved paper properties.

A. TWO-ROLL PRESS ARRANGEMENT

Straight-Through Press

The straight-through press is the oldest and simplest arrangement. The sheet is carried through the press by a felt on its underside. A smooth roll contacts the topside of the web (**Fig. 4.1**). The primary function of this press is dewatering of the web. The secondary functions are to generate topside smoothness and to densify the web.

The straight-through press arrangement requires minimum framework, and is the easiest to operate, maintain, and clothe. Absence of the top framework makes roll changes easier. Felt runs are very simple and require few rolls.

Shortcomings include limited water removal capacity because it is only one-sided, nonuniform treatment of the web, and the requirement for open draw after the first press where paper is still wet.

Sometimes, the felt lead-in roll is raised so that felt and paper touch the top roll well before the nip. This arrangement reduces blowing (air pockets between felt and sheet ahead of the nip) and wad burning, which may result from lumps of paper which during threading refuse to go through the nip and slide on the felt (**Fig 4.2**).

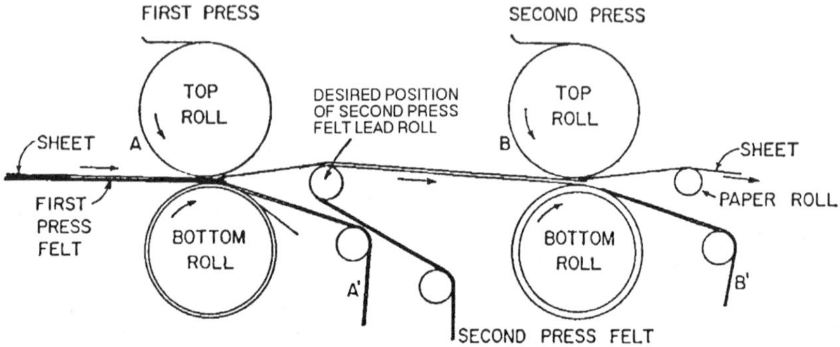

Fig. 4.1: Straight-through press

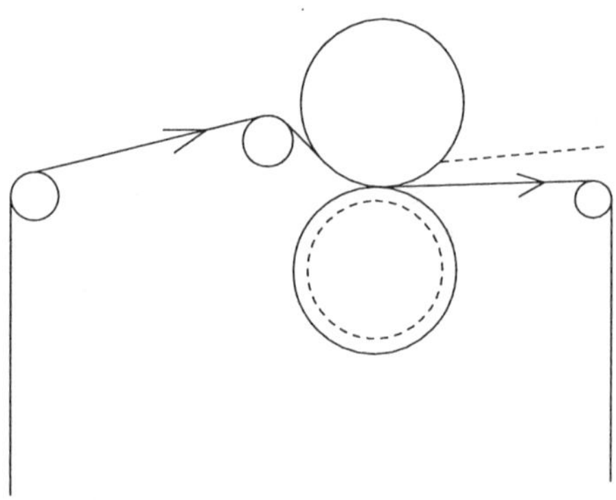

Fig. 4.2: Increased in-going wrap to reduce blowing, often referred to as the Toledo wrap

Reverse Press

A reverse press has basically the same geometry and functions as the straight-through press, except that the sheet passes through the nip in a direction opposite to the machine travel. The bottom side of the sheet now contacts the top smooth roll, therefore reducing bottom side wire mark and improving smoothness (Fig. 4.3).

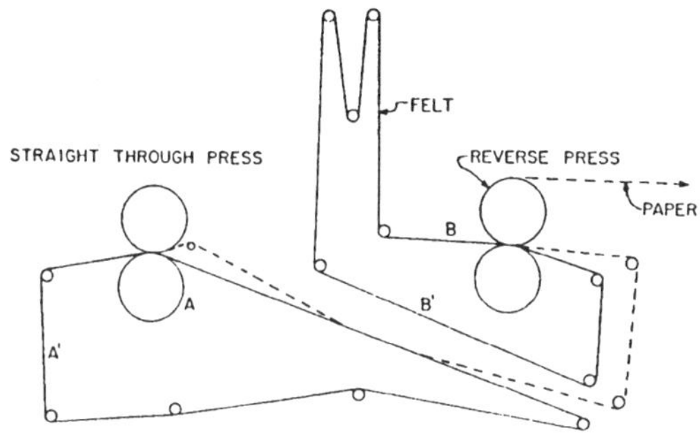

Fig. 4.3: Reverse press

Two-Roll Inclined Press

Sometimes, two press rolls are arranged in an offset pattern rather than on top of each other. Such nips point upward or downward, instead of out being in horizontal direction.

Inverse or Upside-down Press

The inverse or upside-down press has a felted top roll and a smooth bottom roll, generally arranged on top of each other. The top roll may be a suction roll, or it may be grooved or blind drilled, depending on how much water must be removed (Fig. 4.4). Note the open draw after the plain bottom roll.

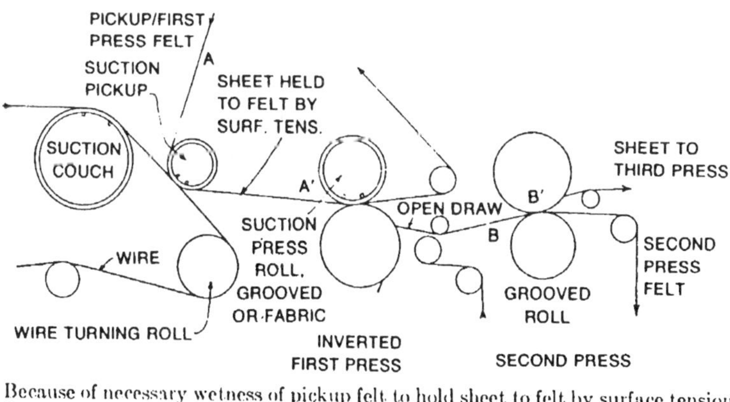

Because of necessary wetness of pickup felt to hold sheet to felt by surface tension first press nip is lightly loaded to avoid sheet crushing. Even then there is some loss of fines and alignment of fibers.

Fig. 4.4: Inverse or upside-down press

If a suction roll is used in the top position, the sheet can follow the top felt and the draw may be closed (**Fig. 4.5**).

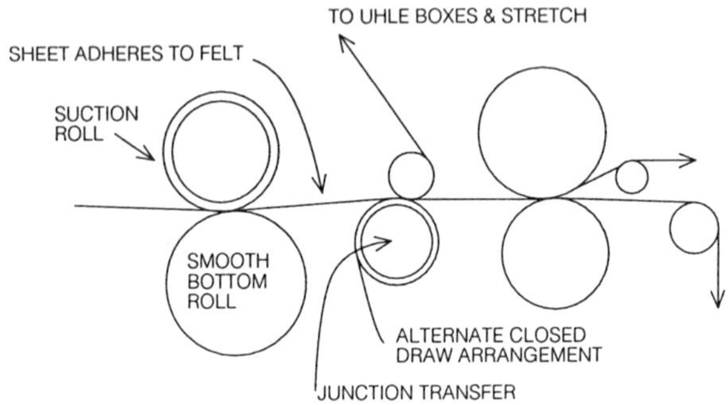

Fig. 4.5: Inverse or upside-down press, with closed draw

Transfer Press

A transfer press may have any of the above configurations and serves the additional function of transferring the web from a top felt to a bottom felt, or of removing the web from the felt (**Fig. 4.6**).

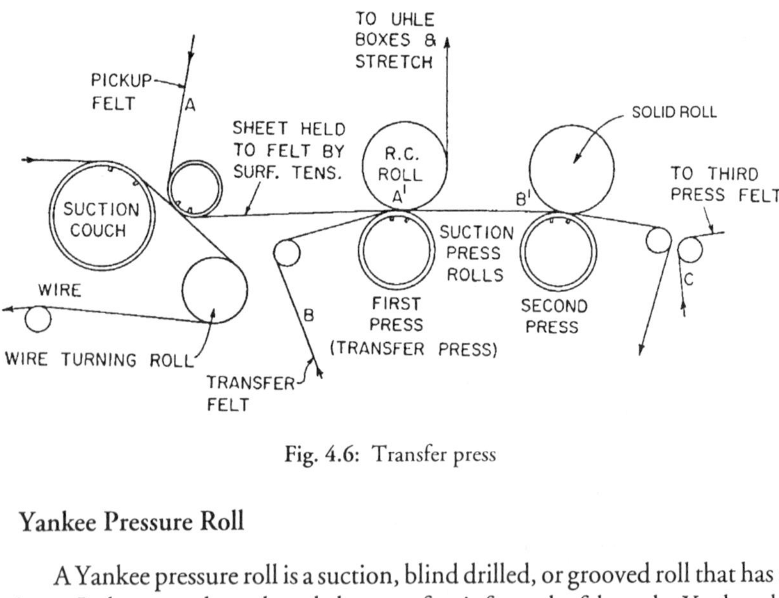

Fig. 4.6: Transfer press

Yankee Pressure Roll

A Yankee pressure roll is a suction, blind drilled, or grooved roll that has two functions. It dewaters the web and also transfers it from the felt to the Yankee dryer surface. Such a nip is usually preceded by one or two single- or double-felted presses.

The web may stay with the top pickup felt until it is transferred to the Yankee dryer surface.

Sometimes, two pressure rolls are used to provide additional dewatering, more intimate contact, and better drying. The felt then leaves the sheet after the first pressure roll, is passed around a felt roll, and then back into the second pressure roll nip.

Often, Yankee tissue machines use only one felt. The first pressure roll must then have a wide suction box, to ensure that the wet tissue follows the felt around the 180° turn. This arrangement is similar to **Fig. 2.8** (p. 15), and is suitable for heavier basis weights.

Combination Suction Pickup First Press

The combination of suction pickup and first press was originally developed in the late 1920s (see **Fig. 2.5**, p. 13). A wide suction box is used in the pickup roll to guide the web into the first nip.

With a conventional suction pickup system, the pickup felt is quite wet in order to hold the web on the underside of the pickup felt. In this combination, however, a dryer felt is desirable to achieve higher water removal without crushing.

The suction pickup roll generally can be held in the fixed position to simplify vacuum connections. For the transfer, the wire is moved toward the suction pickup roll. This arrangement is very compact and reduces the number of felt rolls and the amount of framing normally required for a suction pickup arrangement. Felt changing is simplified.

The bottom roll must have good release properties. Note the open draw after the first press nip.

B. THREE-ROLL PRESS ARRANGEMENTS

Horizontal Three-Roll Press

The three-roll press arrangement (**Fig. 4.7**) is very compact, requiring minimum framework and felt rolls. An open draw transfers the web from the wire to the first press felt. At the first nip, the web follows the center roll to the second nip, where it is transferred to the second felt.

The center roll must have a nonpicking easy-release surface, and it must be doctored. Since two rolls oppose each other, the second nip should have only slightly higher load than the first, typically 50 pli maximum differential. A paper roll is sometimes used following the second nip. This separates the web from the felt in order to decrease rewetting.

After the first nip, the first felt should wrap the center roll a little to ensure that the web does not drop off of the center roll.

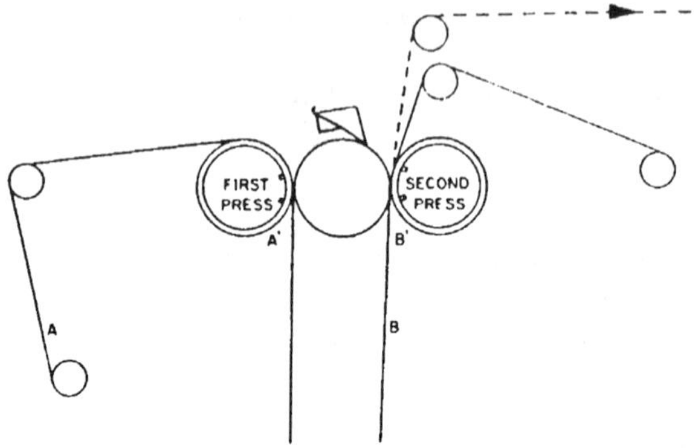

Fig. 4.7: Horizontal three-roll press

Cloverleaf Press

The cloverleaf press (**Fig. 4.8**) is a modification of the horizontal three-roll press, designed to improve sheet runs. Note that the center roll now must resist considerable upward forces. It is crowned for given nip loads, and uniform nips will result only at those loads.

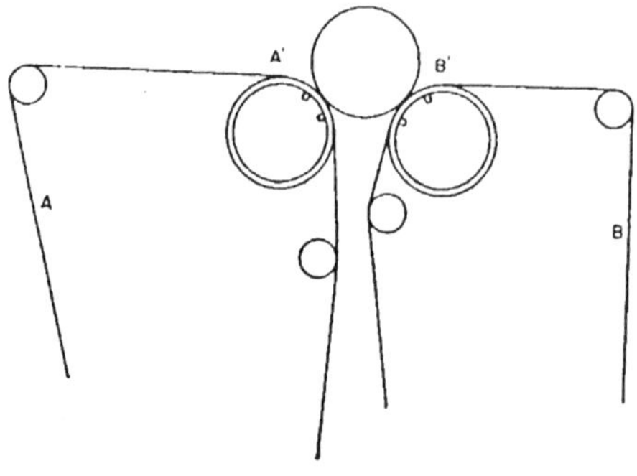

Fig. 4.8: Cloverleaf press

Three-Roll Inclined Press

The three-roll inclined press is another modification of the basic principle (**Fig. 4.9**). This arrangement was developed during the 1950s for use on suction pickup machines to achieve two press nips before the first open draw, primarily for high-speed newsprint and fine paper machines. A suction press roll must be used in the first nip to facilitate easy water removal. The web then follows the center roll to the second nip for additional dewatering. The web is then removed from the center roll and passes via an open draw to the next nip. Note again the limitation of the load differential between first and second nips.

The center roll, which contacts the bottom side of the sheet, must have an easy- release, nonpicking surface. Generally, this is a granite roll or a bone-hard, rubber-covered roll. New composite covers are also being introduced. Often, the third press nip is a straight-through with a bottom felt and a granite roll treating the top surface of the paper.

Unlike the two previously discussed three-roll presses, framework for this arrangement is more complicated and more felt rolls are required. Roll and felt changing is more difficult with this press.

Such presses are operated at speeds up to 3000 ft/min.

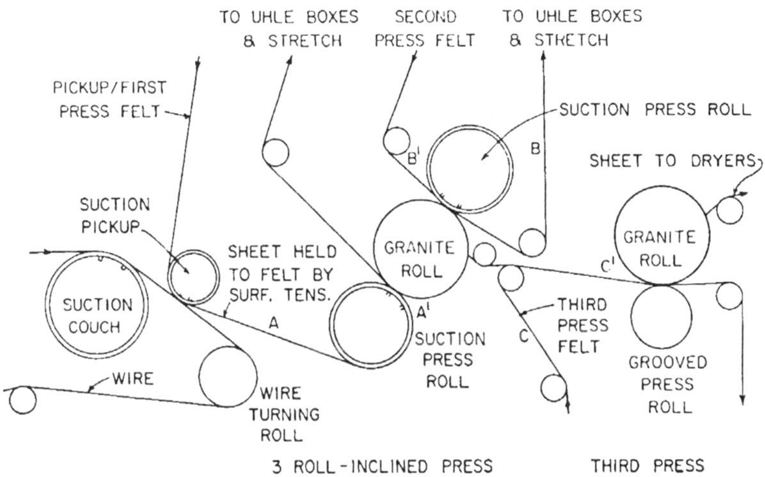

1. Because of necessary wetness of pickup felt to hold sheet to felt by surface tension first press nip is lightly loaded to avoid sheet crushing. Even then there is some loss of fines and realignment of fibers.
2. With light load on first press second press is lightly loaded for load on second is limited to a load differential of 50 lb/inch greater than that on first press.

Fig. 4.9: Three-roll inclined press arrangements

Three-roll Pickup Press

The three-roll pickup press (**Fig. 4.10**) combines the ideas shown in **Fig. 4.9** with a direct suction pickup. This is a very compact press arrangement with overall characteristics very similar to those discussed in the preceding section.

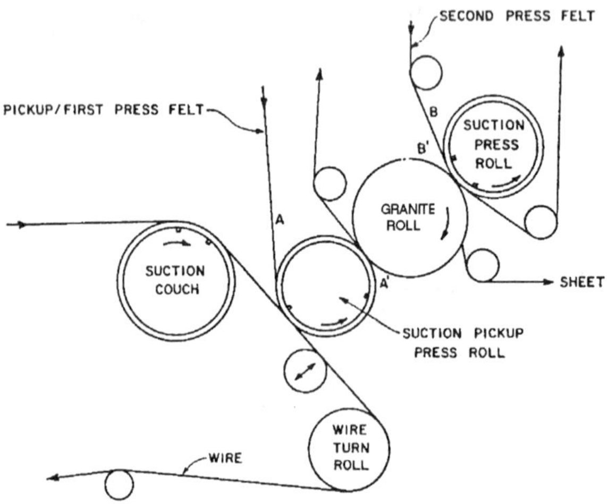

Fig. 4.10: Three-roll pickup press

Double-Felted First Press

For additional dewatering before the sheet meets the first plain roll, a double felted first press may be arranged as shown in **Fig. 4.11**. After the suction pickup, the wet web follows the pickup felt into the first nip, on the felt underside. The bottom side of the paper meets a bottom felt before the first nip.

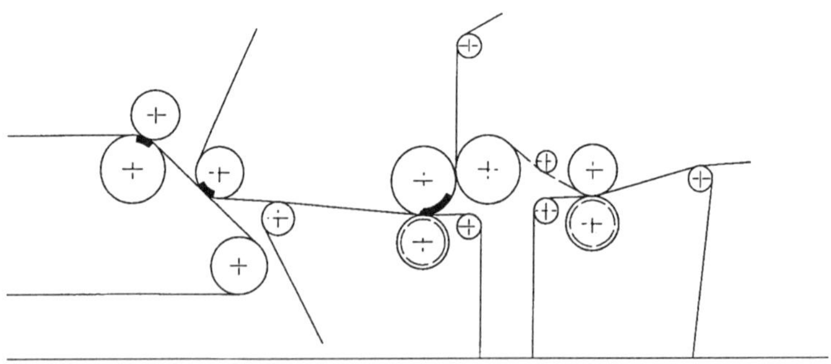

Fig. 4.11: Three-roll arrangement with double-felted first press, forming two consecutive nips. Also shown are suction pickup and single-felted third press.

The paper then follows the pickup felt, aided by the suction roll to the second nip, for transfer to the plain roll which treats the bottom side of the paper.

After release from the third roll, the paper goes on to further pressing.

C. FOUR-ROLL PRESS ARRANGEMENTS

For today's higher speed paper machines, there is increasing need to eliminate open draws, particularly on lightweight papers. This is achieved with four-roll presses.

Four-Roll Three-Nip Press

The now-common multiple nip arrangement (**Fig. 4.12**) features suction pickup, double-felted first press, transfer to a plain roll in the second nip, and pressing of the paper in a third nip on the plain roll - all without open draws. This arrangement may run at 4500 ft/min.

This press arrangement is now widely used, particularly in newsprint machines. In fine paper machines, it is often followed by another press nip with a plain roll touching the top side of the sheet.

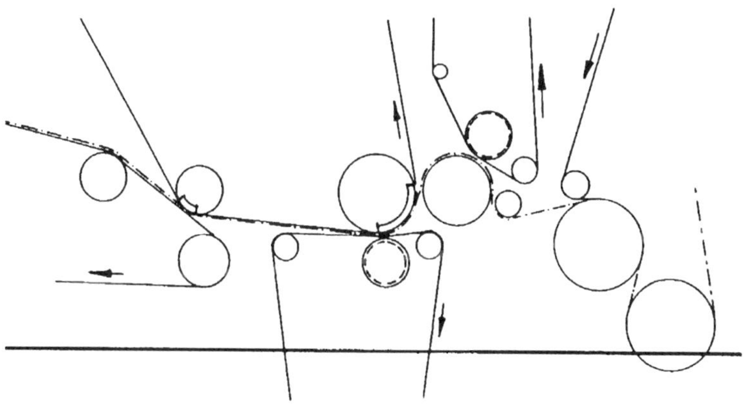

Fig. 4.12: Four-roll three-nip press, in which paper passes through three nips without any draw, always supported by felts and/or roll surfaces.

Four-Roll Press, With Draw

This arrangement uses two press nips before an open draw, with vented nip (grooved) rolls (**Fig. 4.13**). In comparison with the three-nip arrangement discussed above, shadow marking would be reduced because less water is removed through the holes in the suction roll.

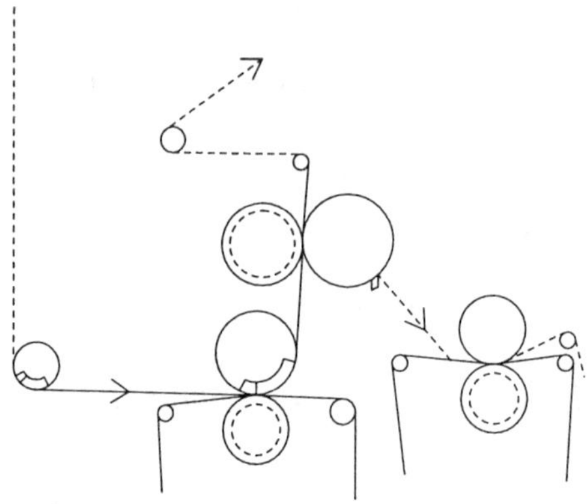

Fig. 4.13: Four-roll separated press arrangement. Third press is also shown.

D. FIVE-ROLL PRESS ARRANGEMENTS

Five-Roll Press, With Draw

Mechanically more-complicated press arrangements are being developed as better felts become available and as greater investment is justified by even small improvements on high-production paper machines.

An example is a five-roll press, with draw (**Fig. 4.14**). The suction pickup is followed by a double-felted first press, wherein the sheet follows the pickup felt to a separate second press nip formed by a felted grooved or blind drilled roll, and a plain roll. The latter carries the sheet to the third nip, which is followed by open draw transfer to another suitable press.

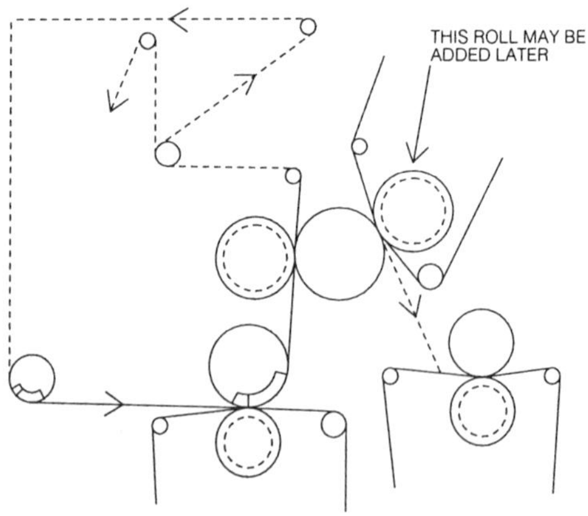

Fig. 4.14: Five-roll separated press arrangement

Separation of the first press and second press nips (in comparison with **Fig. 4.12**) allows higher nip load in the second nip and, therefore, greater dryness of the wet web at the open draw. It also reduces shadow marking tendency due to double pressing while the web is on the suction roll.

E. OTHER PRESS ARRANGEMENTS

Single-Felted Last Press

After the open draw which follows the second or the third press nip, a last press is commonly installed. Its functions are further water removal, surface treatment, and sheet compaction.

An example for a last press is included in **Fig. 4.14**. Note that the plain roll touches that side of the paper which has not previously been in contact with a plain surface.

Last press nips are highly loaded. The felted rolls are grooved or blind drilled. Their roll covers may be stainless steel or rubber, depending on load and mill preference.

Double-Felted Last Presses

For paper grades where maximum dryness after the press is more important than surface finish (i.e., linerboard), the last press may be double felted.

In the 1980s, many highly loaded presses (2,000 pli), featuring "jumbo" rolls up to 80 in. in diameter were installed.

Considering the large roll diameters and the thick felts, nips as long as 4 in. may be generated. For more detail refer to Chapter 3.

In an effort to provide even greater loadings, shoe-type presses have been developed, as described in Chapter 3. Such presses are frequently used in the final pressing position.

Shoe Press Applications

Shoe presses with high pli loadings are generally applied in the last press position, i.e., second or third press. However, there are also instances where a shoe press may be used as a first press (see **Fig. 4.15**).

In order to eliminate an open draw, a bottom felt may be run from a first suction press into a shoe press, as shown in **Fig. 4.16**.

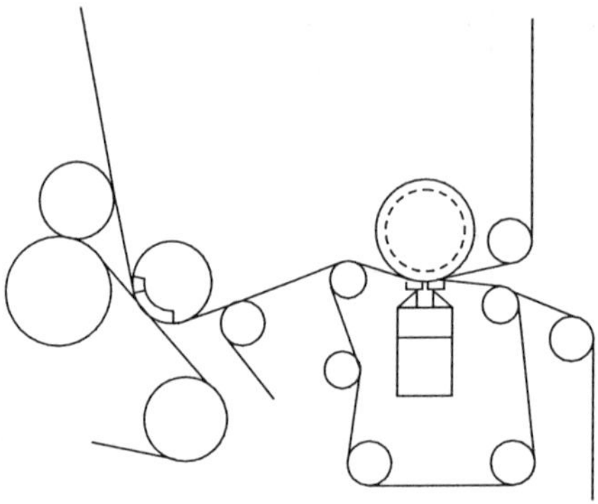

Fig. 4.15: Shoe press in first press position

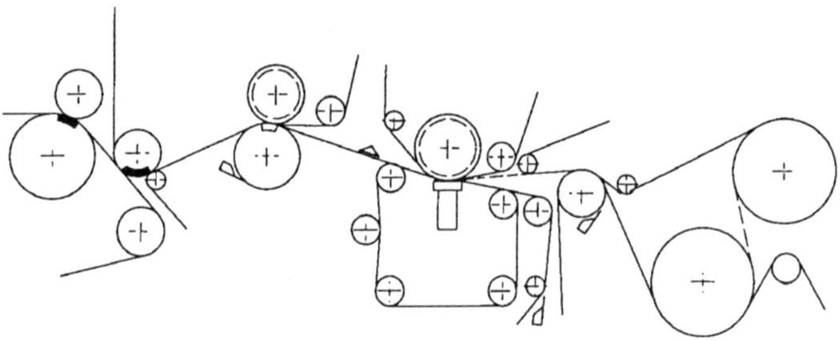

Fig. 4.16: Tandem felt in roll and shoe press. A single bottom felt is used to carry the sheet from near the suction pickup to a first roll press and on to shoe-type extended nip press.

In an effort to increasingly raise the consistency of the sheet before the dryer section, press sections have been installed with two shoe presses in sequence. Examples for this are shown in **Figs. 4.17 and 4.18.**

Hot Pressing

The use of a heated roll which is in contact with the paper has gained increasing acceptance in the paper industry. Felted elastomeric covered rolls are pressed against the heated roll, forming a very effective press nip. One example for this is shown in **Fig. 3.8** (p. 27), which features two press nips on one steam heated cylinder. The surface finish of the paper side which touches the cylinder is very smooth. If both sides of the paper must have equal finish, dual hot presses will be considered in the future. An example for this is shown in **Fig. 4.19**.

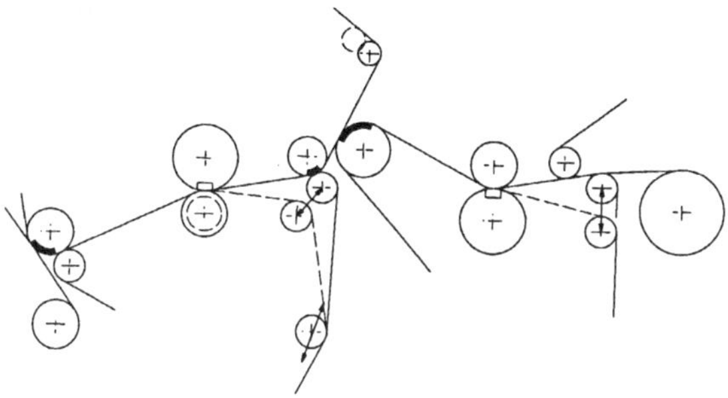

Fig. 4.17: Press section for two shoe-type presses for corrugating medium. Note that the second shoe-press is single felted. Nip loads to 4000 pli and 5000 pli, respectively. The shoes are in an elastomeric flexible shell.

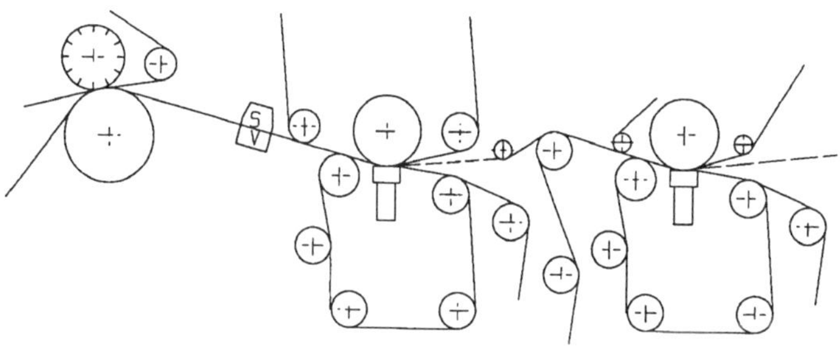

Fig. 4.18: Two shoe-type presses, in sequence. In the example, all presses are double-felted. The first bottom felt carries the paper from the suction press to the second press. There are open draws before and after the third press.

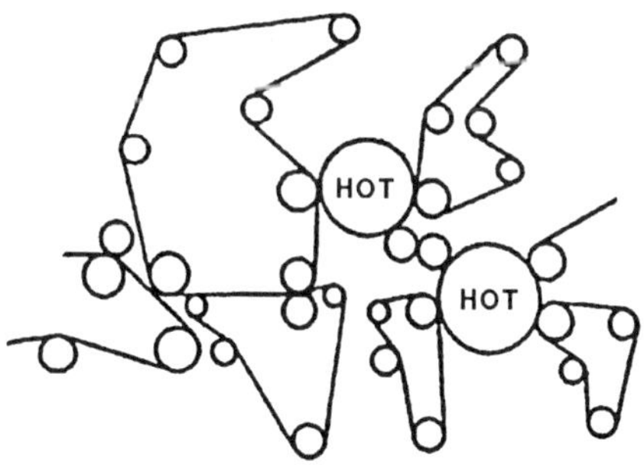

Fig. 4.19: Two hot presses in sequence. Two direct action hot presses may be arranged in sequence to achieve more equal surface finish and to raise consistency to the dryers even further.

Chapter 5

Rolls of the Press Part

The press section of the paper machine can be very simple in design for low-speed specialty machines, or very complex in design for high-speed paper production. The type and design of the rolls of the press part play a very important part in the function of the press.

Press rolls when designated on press section drawings are described with various symbols. Fig. 5.1 illustrates the various press roll symbols typically used on press section outline drawings.

A. PRESS ROLL TYPES

Plain Press Roll

The plain press roll consists of a shell body of bronze, cast iron, stainless steel, or chilled iron, which may or may not be covered. Plain press rolls are also referred to as solid press rolls, even though the shell body is hollow cored. The plain press roll can be covered with rubber, polyurethane, or fiberglass material. The plain press roll is the oldest type of press roll used in the press section. All other types of press rolls were derived from the plain press roll.

Suction Roll

The first major improvement in the design of press rolls was the suction roll, which consisted of a through-drilled bronze shell body fitted with an internal, stationary suction box assembly connected to a vacuum source. The vacuum connection was designed to be from the front or rear of the roll assembly, depending upon the particular requirements.

The principle of the suction roll has not changed since its introduction into the press section to aid in the removal of water from the paper web. The mechanics of actual water removal in the nip have changed with the evolution of press felt design.

Suction roll shells are now manufactured from bronze, aluminum bronze, and stainless steel. The internal components of the roll may be constructed of cast iron, mild steel with resin coatings, or solid stainless steel.

Depending on the application, the suction roll can be operated with or without a cover material. The style of the drilling pattern and amount of open area are dependent upon the application into which the suction roll is to be placed.

TYPE	SYMBOL	COVERS
PLAIN	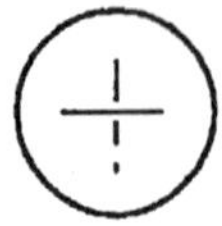	Rubber Polyurethane Bronze Granite Fiberglass
SUCTION		Rubber Bronze Stainless Steel Polyurethane Fiberglass
GROOVED		Rubber Polyurethane Stainless Steel Fiberglass
VARIABLE CROWN		Rubber Polyurethane
BLIND DRILLED		Rubber Polyurethane Fiberglass
COMBINATION GROOVED/BLIND DRILLED		Rubber Polyurethane Fiberglass
ELASTOMERIC		Elastomeric

Fig. 5.1: Press roll types and symbols

Suction rolls are typically used in pickup, top, bottom, multiroll, and transfer positions in the press section. The suction roll is usually a driven roll, with the vacuum discharge taken from the front side of the roll.

Depending upon the application, a suction roll shell has an open area of 15%

to 30%. Drill patterns are carefully selected to avoid stress concentrations, as well as to provide the optimum pattern for dewatering and low noise levels.

The application of the suction roll, grade of paper, speed, and nip load will determine the final construction of the suction roll assembly.

For the machine speed and loading applications, some mills have opted for suction rolls manufactured completely from stainless steel material for maximum strength and corrosion resistance.

When designing for suction press nips, care must be taken to select the proper inlet and outlet geometry of the nip and the type of press felt to avoid shadow marking.

The suction roll has been an integral part of the press section for a number of years. With the advent of jumbo presses and shoe-type presses, it has sometimes been predicted that the suction roll would vanish from the press section. This has proven false, with the suction roll still playing an integral part in the design of modern press sections.

Granite Roll

A granite press roll consists of a heavy-walled, granite roll body, which is machined from quarried blocks of granite. The granite roll body is fitted to steel through-shafts, tie rod assemblies, or journal assemblies to complete the assembly.

The granite roll is a plain press roll that is in direct contact with the wet paper. It imparts smoothness to the side of the paper which it is contacting. Wet paper wants to stick to a smooth surface and considerable pull or tension is needed to peel off the paper. The nonhomogeneous, yet smooth surface releases the paper relatively easily. Granite has found widespread use in the press section, particularly in newsprint production, because of its excellent release properties which are now being duplicated by manmade coverings.

Granite rolls are sensitive to thermal shock, which may cause breakage. Also, granite is not uniform in all directions and it may expand nonuniformly, leading to press vibration. The search for good manmade material is intense, and paper technologists should follow new developments closely because of the promise of significant economies in press operation.

Grooved Press Roll

The grooved press roll was the first variation applied to the plain rubber cover material of a plain press roll. Grooves of a uniform depth and width, and an axial pattern, were cut into plain rubber covers to reduce the hydraulic resistance to flow in the press nip and consequently enhance water removal in the press nip.

Grooving is also used on press rolls with stainless steel covers, polyurethane covers, and fiberglass covers. The pattern of the roll grooving will vary depending upon the roll position, cover hardness, and roll supplier.

Blind Drilled Roll

The blind drilled press roll was the next step after the grooved cover in the evolution of rubber cover design. The principle of suction roll drilling was applied to a solid roll cover, where the solid cover is drilled to a partial depth to provide open area similar to a suction roll. The open area of a blind drilled roll is greater than that of a grooved roll.

Combination Grooved/Blind Drilled Roll

After the grooved and blind drilled covers became accepted cover designs, the two drilling patterns were applied to the same press roll cover to obtain the maximum benefit of open area from both designs. This design has proven successful in several last-bottom positions of highly loaded press positions.

Self Peeler Roll

Self peeler press rolls are plain press rolls, supplied with a rubber cover material that is designed to enhance the release of the paper web from the roll. This design is used as an alternative to the granite press roll.

Steam Heated Roll

The steam heated press roll is a plain press roll, designed as a pressure vessel to accept internal steam pressure as well as an external nip load force. The steam is used to increase the roll and cover temperature in an effort to transfer this heat to the sheet to improve the pressing action in the nip. Increasing the sheet temperature in the press nip lowers the viscosity of the water in the sheet and greatly improves the rate of water removal from the sheet. The steam heated roll can be covered with a temperature resistant, rubber cover material, or a special sprayed-on release material.

Shoe With Belt Type Press

A departure from the use of rolls in the press section to create a nipping action to remove water from the sheet has led to the development of the shoe-type press. The shoe-type press consists of a top plain press roll and a bottom, hydraulically actuated shoe surrounded by an impervious belt (Fig. 5.2). This is typically a double-felted press, designed to provide a long nip effect at pressure levels far higher than those achievable with conventional press rolls.

A shoe with belt type press can develop a nip that is up to seven times wider than a nip using conventional press rolls.

Nip loads up to 6000 pli have been achieved with the shoe with belt type presses to achieve 10-in. wide nips.

The greatest success to date for this press style has been on heavyweight grades such as linerboard and corrugating medium, where high volumes of water must be removed in the press nips at high speeds.

Shoe With Elastomeric Sleeve Roll

In a development similar to the shoe with belt type press, an elastomeric-covered press roll, consisting of a fixed shaft, rotating shell, hydraulic shoe, and flexible rotating sleeve or blanket, was conceived. (See **Fig. 5.2**)

Elastomeric covers have physical properties that permit the cover to deform under the great pressure of the press nip. This creates a wide nip, longer dwell time, and increased water removal capability, similar to the shoe with belt type press.

The cover design is proprietary and varies with the supplier of the roll and cover. These covers represent some of the more advanced developments in roll coverings and roll design.

Typically, the elastomeric-covered roll will operate against a solid rubber covered roll as a double-felted press.

B. ROLL CROWNING

The rolls of the press must be crowned in order to obtain uniform pressing and sheet profiles from press nips. Roll crowning is the intentional placement of a profile into the outside diameter of a press roll to compensate for the deflection of the press roll, due to externally applied nip forces and its own weight.

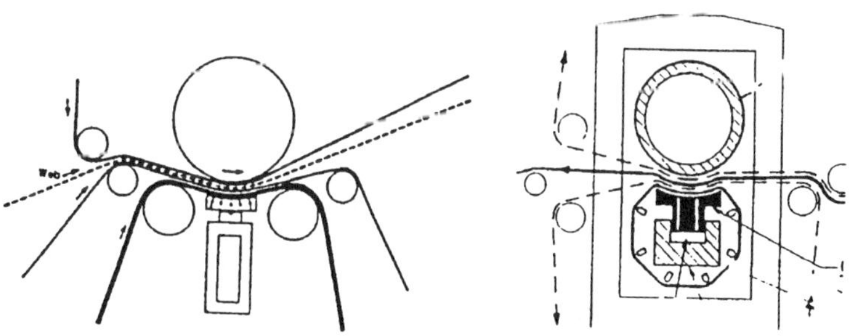

Fig. 5.2: Long-nip presses: shoe type with belt (left) and shoe type with elastomeric bottom roll (right).

The crown of a press roll is a designed contour on the surface of the roll and is continuously decreasing nonlinearly from the center point to both ends.

A perfect crown for a specific loading will give equal pressure across the face of the two mating rolls. **Fig. 5.3** depicts nip impressions taken under proper crown, excessive crown, under crown, and irregular crown conditions.

Fixed Crown Roll

A fixed crown roll refers to any press roll which is designed to operate at a single nip pressure with a specified amount of crown in the roll. To operate at a different pressure, the roll must be recrowned for the new nip pressure. Plain body and suction rolls fall into the category of fixed crown rolls.

Variable Crown Roll

Variable crown rolls are plain body-type press rolls, designed with an internal dead-through shaft which supports the outside roll body. The internal shaft is designed to be cambered by a hydraulic system to allow the roll to vary its profile in response to different externally applied nip forces.

The first variable crown press roll consisted of a single internal pressure zone to effect the camber of the roll. This roll design is still in wide use in press sections where nip pressures are varied greatly in efforts to produce desired sheet properties from the press.

Variable crown rolls can be covered with any style of material. The roll is typically a driven roll with a remotely located hydraulic system dedicated to the control of the press roll.

Zone Control Roll

The zone control roll is the further development of the variable crown roll. The zone control roll is hydraulically actuated like the variable crown, but the roll's internal pressure zone is broken into several separate zones, each of which can be individually actuated.

The individual zones provide more exacting control in the ability to profile the press roll in response to the applied nip forces.

Several versions of the zone control roll are available for use in the press part. Construction details differ, but all produce the same effect.

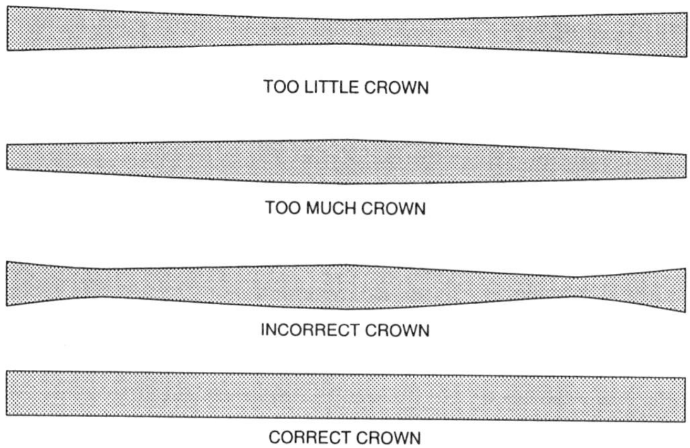

Fig. 5.3: Nip impressions. These nip impressions should be taken when rolls go into operation and also when they are removed for regrinding. From these impressions it is possible to adjust the amount of crown to provide a more uniform nip.

Zone control rolls can be covered with any type of material. Zone control rolls are typically driven and supplied with a dedicated hydraulic control system.

C. CARE AND MAINTENANCE

Periodically, the press rolls must be reground. Frequency is determined by experience gained during operation of each particular press roll. Rubber covers wear to the extent that proper contour is lost, become out-of-round or corrugated, or develop surface damage or cracks.

Corrugation, ridging, or out-of-roundness is often due to misalignment of rolls. When suction rolls are removed for regrinding and replaced with a spare, the suction boxes are removed for cleaning, inspection, and replacement of seals if required. The inside surface of the shell is inspected for scoring, and suction holes are cleaned. The practice in most mills is to drill out the hard deposits which collect and fill the suction holes. It is possible to remove the deposits which collect in suction roll holes at some mills by soaking the shell in a prepared cleaning solution or by use of external high pressure needle cleaning showers.

Felts can be damaged and their life reduced, or poor press operation can result from the following: improper roll crowns; improper density and thickness of rubber cover; operating rolls with loose covers or covers that have checks, corrugations, or cracks; uneven or improper loading; failure to release loading and to separate rolls when at rest; misalignment of rolls; unbalance of rolls; use of solvents, oils, and chemicals injurious to the rubber covering; and operation of a suction or grooved roll with plugged holes or grooves.

When removing and storing press rolls, the following points of care should be exercised:

a. Use slings which will distribute loading over a large area of the roll face. Roll and sling surfaces must be clean, smooth, and free of foreign particles. Use rope or synthetic strap slings only on journals.
b. When the machine is shut down, rolls must be separated and out of contact.
d. To reduce oxidation, cracking, and checking of roll surfaces, rolls should be stored in a cool, dry area and out of direct sunlight. The rolls should be wrapped in paper or placed in a wooden box.
e. Support rolls only on journals - never on the roll surface. Rotate rolls in storage periodically.
f. Keep roll surfaces free of grease, oil, and harmful chemicals.
g. Regrind rolls that have been in storage a considerable time before installing on a machine.

D. OTHER ROLLS IN THE PRESS PART

Felt Rolls

Rubber-covered felt rolls are used in the press part to convey the press felts through the press nips, and through the conditioning, guiding, and tensioning stages. Felt rolls must be designed to run the felts without excessive deflection, which can affect press performance.

Press felt rolls are typically supplied with plain covers, with wormed-type covers supplied for the stretch roll positions to provide a spreading action to the felt. Wormed covers are typically .0625 in. to .125 in. larger on the diameter than plain felt roll covers.

The cover material for press felt rolls can be hard rubber, polyurethane, or fiberglass.

Paper Rolls

Paper rolls are used in the press part where open sheet draws exist between press nips, or at the exit of the press into the dryer section.

Paper rolls typically have plain covers, designed for easy release. Cover materials can be Teflon or plasma spray. Typically, the paper rolls are driven to aid in the initial threading of the sheet through the press section.

Spreader Rolls

Spreader rolls are often installed in a felt run to widen and open the felt just before it enters the press nip. Spreader rolls are made up of several short lengths of straight rollers, or several narrow rollers covered by a flexible tubing mounted on a bent or bowed shaft.

Rather than a concave or convex felt roll, a spreader roll can be employed to maintain a straight felt seam by positioning the bow at right angles to the felt travel.

Suction Felt Rolls

The suction felt roll is a smaller version of the suction press roll, much simpler and lighter in design. Its primary function is to assure paper web transfer from one press felt to another. The suction felt is driven by the felt run and may be mounted on a pivoting mechanism to allow adjustment of the felt runs.

Please refer to **Fig. 5.1**, which depicts the common symbols used to designate the various roll types used in the press part.

E. PRESS NIPS

Dwell time in the press nip is an important factor in press design. The selection of the style of press roll could have a dramatic effect on operating nip width.

Figure 5.4 provides a comparison of the resultant nip widths from conventional presses, jumbo presses, and long nip presses.

Conventional presses are those presses utilizing roll diameters up to approximately 72 in. with suction rolls, plain rolls, blind drilled, granite, or grooved rolls. Nip widths up to 40 mm are obtainable with conventional press rolls. Jumbo presses are defined as those presses with roll diameters up to 72 in., which can utilize plain rolls or blind drilled rolls. Nip widths up to 100 mm are obtainable with jumbo press rolls.

Long nip presses are the shoe type with belt presses and elastomeric covered presses. Nip widths up to 250 mm are obtainable with long nip presses. Jumbo presses can create press nips which are 2.5 times as wide as conventional presses, while long nip presses can create press nips seven times wider than conventional press nips.

Press nip selection is a function of sheet grade and properties and speed. For slower-speed machines, conventional nip presses deliver enough dwell time in the nip for satisfactory dewatering of the sheet.

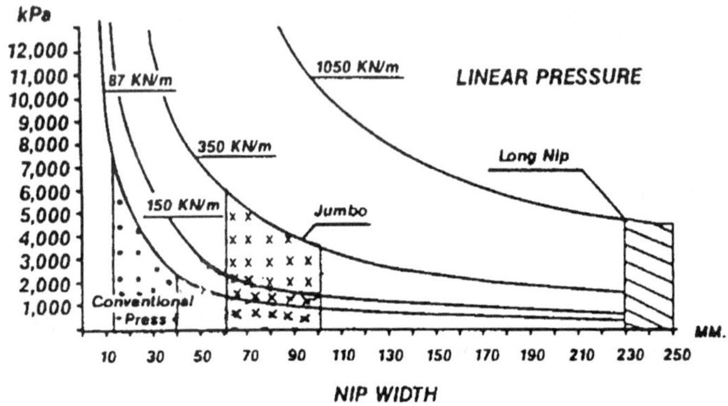

Fig. 5.4: Comparison of various types of press nips

As speeds increase and reduced nip dwell time becomes a critical factor, the jumbo and long nip presses become the effective methods of maintaining nip dwell time for optimum press performance. Based on the press nip data from **Fig. 5.4**, the maximum pli loadings for the major press categories are presented in **Table 5.1**.

Table 5.1 Maximum pli loadings for major press categories.

Press Category	Maximum PLI Loading
Suction rolls	800
Plain press rolls	1600
Jumbo press rolls	2500
Shoe and belt type presses	6000

As press roll technology advances, so will the operational limitations of the various roll designs. Press roll design must stay in step with the increasing demands for quality paper products, produced at the highest attainable machine speeds.

Chapter 6

Press Fabrics

A. FUNCTION

The cost of dewatering the sheet in the press section is significantly lower than that of dewatering in the dryer section. For this reason, the value of efficient press fabric performance cannot be overemphasized. The primary function of a press fabric is to provide resilient, permeable support for the sheet in the nip to maximize water removal. The press fabric provides the necessary cushion and void volume to receive water expressed from the sheet. The fabric also conveys the wet sheet through the press section and provides power transmission to turn rolls in the fabric run.

In serving these functions, the fabric must not have an adverse effect on sheet properties. The fabric must provide pressure uniformity to prevent sheet marking from the press roll, base fabric, and fabric surface. To provide economical operation, the fabric must have adequate water handling characteristics over its life span. It must be durable and resistant to compaction, chemical degradation, and the tendency to fill up with furnish particles. It also must provide good machine runnability with freedom from sheet transfer, fabric guiding, vibration, and other problems that reduce efficiency. To satisfy these requirements, the fabric manufacturers provide a wide variety of design constructions. This custom engineering is necessary due to the various combinations of paper grades, machine speeds, press section arrangements, fabric conditioning equipment, and so forth found in the paper industry.

B. MANUFACTURE AND DESIGN CONSTRUCTION

Materials

The primary materials used in the manufacture of press fabrics are polyamides; nylon is a common tradename. Polyamide fibers and yarns are purchased in various sizes and with different specifications for properties such as tensile strength and degree of preshrinkage. The specifications chosen depend upon the end use. Polyamide fiber is durable, strong, and resistant to wear. For these reasons, it rapidly replaced wool in most press fabric applications.

Some applications utilize wool or high-melting-point synthetic fibers in the paper-side surface to minimize localized surface melting (wad burning), which occurs when the sheet breaks and builds up in front of a press nip. A few isolated applications utilize polyester for its superior resistance to stretch.

A press fabric is made from yarn used in weaving and batt fiber used in needling. Typical yarns fall into four categories:

1. Spun yarns made by twisting short staple fibers together
2. Multifilament yarns made by twisting continuous filaments together
3. Treated yarns which are typically chemically treated multifilaments altered by heat curing
4. Monofilament yarns which can be used singularly or in a cabled, or plied, arrangement.

Needle batt fiber is short staple fiber typically about 3 in. in length and is available in a variety of sizes, degrees of preshrinkage, degrees of crimp, and so forth. The most common term associated with needle batt is *denier*, which is the weight in grams of 9000 meters of fiber length. The metric unit *decitex* is the weight in grams of 10,000 meters of fiber length. Typical denier sizes used in press fabrics range from 1.5 to 80 corresponding to fiber diameters from 0.014 mm to 0.100 mm, as shown in **Table 6.1**.

Major Processing Stages

Figure 6.1 shows the major processing stages of press fabric manufacturing. Yarns are woven together to form a base fabric. This woven base fabric can be made in a variety of weaving patterns; can use different yarns; can be single layer, multilayer, or a combination of two or more independently formed base fabrics; and can be formed either as an endless loop or as a flat piece. Flat woven base fabrics can be seamed and made endless prior to needling, or they can be made with loops on each end that will be pin-seamed (joined) together at time of installation on the paper machine.

Table 6.1 Polyamide batt fiber

Denier	Decitex	Diameter
1.5	1.7	0.014
3	3.3	0.019
6	6.7	0.027
15	16.7	0.043
25	27.8	0.056
40	44.4	0.070
80	88.9	0.100

Denier is a textile term equal to grams per 9000 meters of fiber length.
Decitex is grams per 10,000 meters of length.
Diameter of the fiber, for a round cross-section, is shown in millimeters.

Batt fiber is carded, or combed, in order to orient the fibers in the same direction and to make the mass distribution uniform. Different methods are available for carding and subsequent preparation of the fiber for needling. The two major methods are cross-lapping (conventional needling) and spiraling. In cross-lapping, the batt fibers are layered back and forth in order to form the desired batt weight. As shown in **Fig. 6.2**, when the batt is needled to the base a batt splice is formed where the start and end points meet. The fibers are oriented predominately in the cross-machine direction (CD). In spiraling, the fibers are oriented predominately in the machine direction (MD). The endless spiral is formed either directly on the base fabric or in a separate process.

In the needling process, the batt fibers are driven mechanically into the base fabric through the use of specially designed felting needles. These barbed needles are mounted in reciprocating plates that alternately drive batt fiber down into the base fabric and then retract, leaving the fiber in place. An infinite number of needling techniques can be designed to customize the process.

In the finishing stage, the press fabric, which is now a base fabric with needle batt fiber applied to the paper side or both sides, is washed, dried, chemically treated if desired, and heat stabilized. Heatsetting stabilizes the fabric by altering the orientation of the polymer structure.

Base Fabric

Base fabric constructions are extremely varied. When discussing base fabrics it is common to talk about the number of layers, which refers to the "layers" of machine-direction yarns in the weave pattern. For example, when looking at a cross-section of the felt, a double layer would have two sets of machine-direction yarns, one set stacked on top of the other. **Figure 6.3** shows schematics for some typical base fabrics. The base fabric provides void volume within the structure to accept water expressed from the sheet. It also provides resistance to compaction. An endless woven base is actually woven as a tube. Each edge of the tube is under high tension and must be carefully controlled so that its characteristics are uniform with the rest of the fabric (**Fig. 6.4**). A too-open or too-closed loom edge (sometimes called a turnaround) could cause water handling problems in that area.

Presently in the United States about 90% of press fabric bases are made in endless form, although the use of pin seams has grown in specific applications. The current developments in pin-seamed fabrics are directed toward making the seam area more uniform with the body of the fabric and more durable.

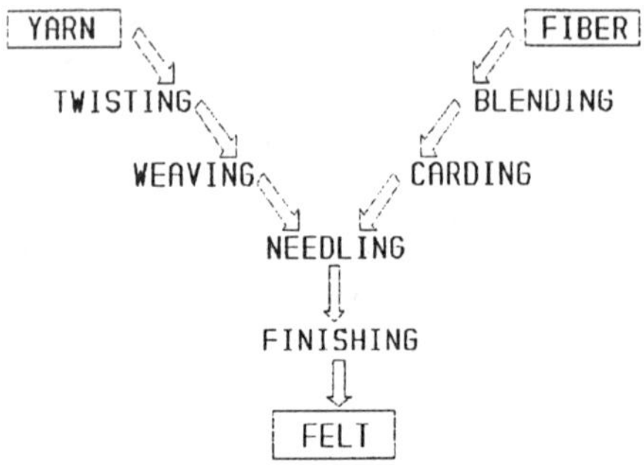

Fig. 6.1: Major processes in press fabric manufacturing

Needled Batt

Batt fiber prevents marking of the sheet by the coarse base fabric, provides more uniform pressure and surface density characteristics to the sheet, provides compressibility in the structure, affects the flow of water from the sheet into the fabric, and protects the base fabric from wear. Customized designs encompass combinations of batt weight, fiber type, and distribution. The fiber can also be oriented to enhance certain properties.

Treatment

Chemical treatments can improve certain fabric properties. They consist essentially of two types: (1) those that internally modify the molecular structure of the materials, and (2) those that supply an external coating on the fibers to protect them and bond them together. Various treatments may be used alone or in combination, thus providing benefits such as reduced fiber shedding, improved resistance to bacterial or chemical degradation, increased resilience, reduced tendency to fill up, increased wear resistance, improved start-up performance, and improved stability. Precompaction treatments make use of pressure and temperature to reduce the bulk of the fabric so that break-in time is reduced. They can be applied with or without chemicals. Precompaction reduces both caliper and permeability, and it has a smoothing effect on the surface of the press fabric.

C. GENERAL PROPERTIES

A number of properties are measured in order to determine a fabric's suitability for a particular application. While fabric manufacturers do not use a standardized set of test techniques, most do some type of testing of the properties that follow.

Unit Weight

The unit weight of the felt, usually expressed in ounces per square foot (oz/ft^2) or grams per square meter (g/m^2) is an important design consideration. The weight influences water handling capabilities as well as compressibility.

Stiffness

A measurement of stiffness is important for assessing the difficulty of installation on machines with small clearances or particularly difficult fabric runs for draping the fabric into the press section. Newer machines are designed to accommodate stiffer fabrics but machines with poor alignment or poor guiding systems often have difficulty guiding very stiff fabrics. Stiffness is measured in the MD and CD; values typically range from 5 g to 50 g (Gurley Stiffness Tester).

Void Volume

Void volume is that space formed in the weave pattern and within the batt fiber structure that can accept water removed from the sheet. In most press nips, water is seen exiting the nip, indicating that the fabric is saturated at midnip. At this point, all of the voids are filled. Sufficient void volume is particularly important when water loads are very high or when hydraulic-related problems such as sheet crushing, shadow marking, and vibration are present. Void volume changes as the felt compacts with age and becomes filled with contaminants.

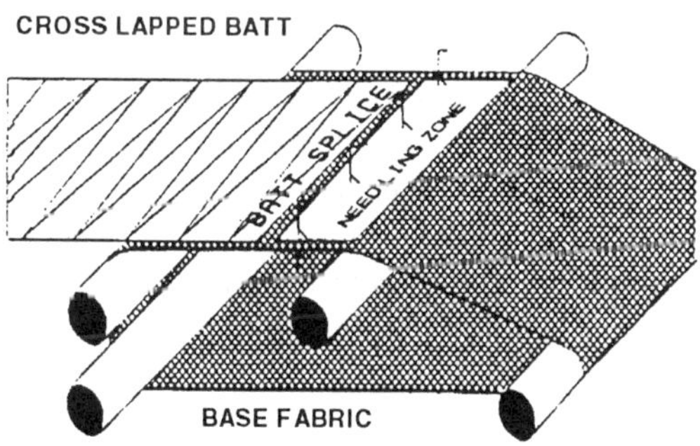

Fig. 6.2: Conventional needling

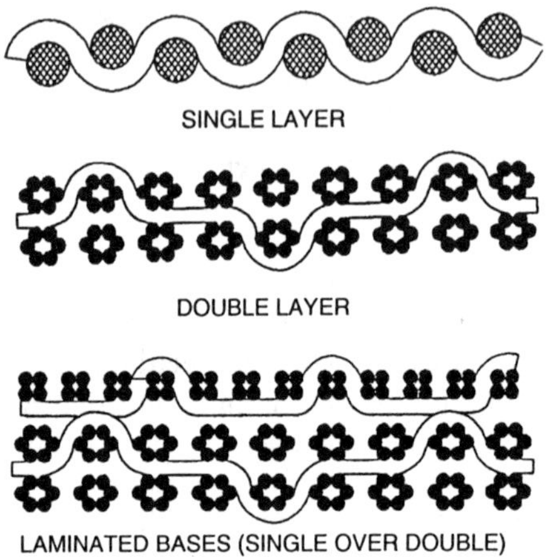

Fig. 6.3: Schematics for typical base fabrics

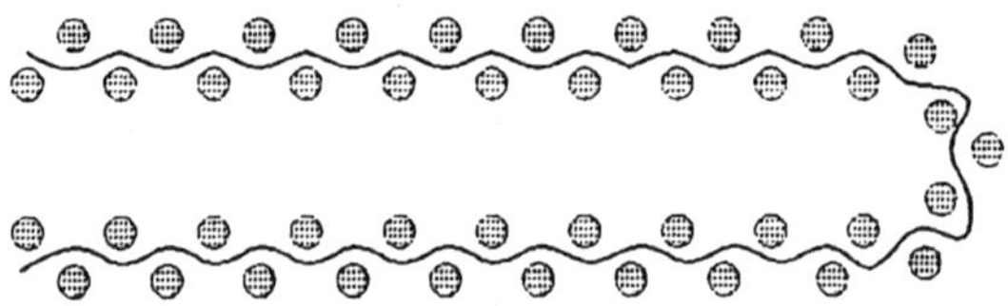

Fig. 6.4: Portion of endless weaving showing right-side loom edge

Compressibility, Hardness

A more compressible fabric has a dampening effect on vibration and will also drive more water into the roll system, provided it is not at its limit. This may be desirable if fabric suction box capacity is inadequate. A more compressible fabric has less hardness which on pressure controlled grades, typically lighter-weight sheets, lowers the effective pressure in the nip and has an adverse effect on dewatering. Measurement of compressibility is not standardized within the industry.

Permeability

Much emphasis has been placed on air permeability measurements taken on new, uncompressed fabrics. This is unfortunate because air permeability measurements were designed to be a quality control tool in the fabric mill. There is no correlation between air permeability values and machine performance. Similar measurements can be obtained on drastically different design constructions. This is best exemplified by the fact that a fabric used inside out is likely to fail although the air permeability is identical to that of the fabric used properly. A more meaningful assessment is that of water permeability measured under load, as this more accurately considers the function of the fabric and the operating conditions. Even this property must be considered in combination with others to determine a design's suitability for a given application. TAPPI Technical Information (TI) Sheet "Characterization of Wet Felts" provides details on air permeability measurement technique.

Permeability measurement of a press fabric in operation can give valuable insight as to how the fabric changes over time. Historical data on a particular design can provide guidelines for scheduling fabric cleaning or removal. TAPPI TI Sheet "Porosity Measurement of Press Felts on the Paper Machine (HSPT)" provides details of this technique.

Caliper

Caliper is easily measured both in the fabric mill and on the paper machine. These thickness measurements give a general indication of the amount of void volume present and how it changes over time. Precompaction treatments reduce caliper. Many mills plot caliper readings versus time to determine when to schedule a fabric change. Two TI Sheets issued by TAPPI address these measurements: "Method for Evaluating Thickness of Wet Press Fabrics" and "Press Section Monitoring".

Tensile and Elongation

The tensile strength of a fabric is substantially greater than the operating tension. The measurement of tensile strength can be of interest when degradation is suspected and both new and previous values can be compared. Elongation characteristics are significant when stretch roll capability is limited or when a fabric goes through two presses and could be subject to stretching due to drive differentials.

Surface Uniformity

The desire for improved sheet surface quality and increased water removal has led to greater concern for fabric surface uniformity. A number of visual tests using gels or other pressure sensitive materials exist; they are subjective in nature and usually rely on comparisons to filed standards. More recently, advanced image analysis techniques have come into use to scan impressions made of fabric surfaces under prescribed loads. The data are then analyzed by computer applications so that values can be assigned that indicate the degree of contact. Surface uniformity affects sheet dewatering and sheet rewet, as well as sheet quality with respect to coating holdout, sheet two-sidedness and so forth. Also, use of higher nip pressures has increased the attention given to this important fabric characteristic.

Fiber Locking, Abrasion Resistance

The ability of a fabric to withstand wear is important for maintaining uniform performance over its life. In some grades, particularly coated ones, loss of batt fiber from the fabric into the sheet can cause operating and quality problems. To test for resistance to fiber shedding and overall resistance to wear, various techniques have been designed to measure the batt loss when the fabric is exposed to a wear condition.

D. APPLICATION

The proper application of a press fabric is critical to sheet quality and to the efficient runnability of the paper machine. Predictable, repeatable performance is important because it allows for maintenance to be planned for times when the machine is scheduled to be down for a fabric change.

Design Considerations

Every fabric position has unique requirements, so there are numerous considerations in selection of the proper design. The broad concerns are for finish, water handling, durability, and runnability. Specific information is necessary for such items as grades produced, furnish, machine type, press loadings, press roll size and type, fabric conditioning equipment, speed, and any particular operating concerns. To achieve optimum performance, the designer must satisfy the requirements by selecting the proper combination of yarn type, weave pattern, batt fiber type, batt distribution, batt orientation, weight of components, needling technique, treatment applications, and heatsetting procedures. After a trial has been evaluated, an analysis of a returned sample can provide additional information for further design refinement.

Effect on Paper Properties

Press fabrics play a major role in machine efficiency and in sheet quality. Uniform pressing with adequate flow characteristics can eliminate negative hydraulic effects such as shadow marking from the roll or press fabric, barring in the sheet due to vibration, and isolated crushing. Sheet surface uniformity can be improved through more uniform pressure imparted by the press fabric in the nip. While two-sidedness is controlled primarily by press geometry, the degree of sheet dewatering in a particular direction is partially influenced by the fabric design used, thereby affecting two-sidedness. Greater sheet strength can be developed through more intensive pressing. This can be influenced by the hardness of the fabric design, which contributes to the psi developed in the nip.

Economics

Press fabrics represent a very small percentage of the overall cost to produce paper but have a large effect on machine performance and, therefore, profitability. Improved press fabric performance can reduce steam usage, thereby providing large energy savings. Alternately, machine speed can be increased, thereby providing incremental tonnage for sale. Incremental tonnage can also be gained by reducing breaks due to poor press performance. Studies show that relatively small increases in speed or small reductions in downtime that result from breaks can add significantly to profits. These gains are often multiples of the cost of the press fabric. Because fabrics have an effect on sheet quality, better performance in this area can also lead to increased paper sales through greater customer satisfaction.

The old measure of life in number of days no longer adequately describes fabric economics. Many mills now schedule fabrics off while they are still performing well. In this way, maintenance activities can be scheduled, risk of machine damage is minimized, poor quality paper is avoided, and operating costs (e.g., steam) are kept in control. For these reasons, fabric cost per ton by itself is an incomplete measure of performance.

E. OPERATION

Press fabric and press section performance can be improved by following certain guidelines and by understanding the causes of and typical solutions to operating problems.

Installation and Start-up

Before a new fabric is unwrapped, the machine position should be thoroughly cleaned. All suction boxes, showers, roll faces, and conditioning equipment should be checked for proper operating conditions and cleanliness. All journals, bearing housings, or other greasy or oily areas that may come into contact with the fabric should be cleaned, and all sharp surfaces should be covered to minimize the possibility of snagging or tearing the fabric.

When the fabric is being installed, the paper side must be oriented properly and the trademark lines should be correct for the direction of travel. The fabric should be carefully worked over the rolls of the fabric run, and it should not be overstretched. When the fabric is in position, it should be spread out to its full width on the rolls. Seamed fabrics are usually equipped with a leader to tie into ropes or into the old fabric so that the new fabric can be pulled through the run. Various types of seam assists are used to hold the seam in place so that the pintle can be put through the loops in order to join the ends. Any grease or oil that gets on the fabric should be removed before the fabric is wet up.

The tradeline should be straight before wetting the fabric. The stretch roll should be adjusted to give sufficient tension to turn all rolls. The press rolls should be very lightly loaded and the fabric rotated slowly for several revolutions. The fabric should be uniformly wet up across its width, which is best accomplished with a full-width shower. During this wetting process, the fabric guide also should be checked for proper operation.

As the fabric is wet up, the stretch roll should be adjusted to maintain uniform tension and prevent wrinkling. The tradeline should be kept straight and the fabric should be operated near the recommended running length provided by the supplier. The fabric should not be trimmed until it has been rotated under full running tension with the press loaded, and the width has stabilized. A sharp knife should be used to trim the edge opposite the fabric guide.

Shutdown Procedures

During brief shutdowns, the fabric should be kept flat under uniform tension. Tension should be relieved during extended shutdowns. When a machine is scheduled for a shutdown exceeding a few hours, it is important to remove contaminants from the fabric with a thorough washing. The fabric should not be roped up for washing. If allowed to remain in the felt, contaminants harden and become difficult to remove later, thereby affecting start-up and subsequent operation. Some mills find it advantageous to apply fabric softener to the press fabric before shutting down. The softener is then rinsed out during wetting up at restart. The fabric, all rolls, and all fabric conditioning equipment should be examined, cleaned, and repaired while the machine is down. At restart, the same start-up procedures should be followed. It is most important to keep the fabric flat, wet it uniformly, and apply the proper tension in order to avoid wrinkling or roping on start-up.

Guiding, Tradeline Distortion

The basic principle for guiding is that when a flat material, such as a fabric under tension, contacts a rotating cylindrical roller, the material will move at a right angle to the roller axis. To be effective, pressure is needed to prevent slippage. This pressure is provided by fabric tension and by the angle of wrap on the guide roll. The wrap is typically 25-35°, divided as evenly as possible on each side of the roll. It is also necessary to have a fixed roll relatively close to the guide roll so that the fabric contacts this roll while it is in its guided position. The wrap on this lead-out roll is normally at least 25° in order to hold the fabric movement. The guide is usually positioned between two fabric rolls so that two-thirds of the distance is from the lead-in fabric roll to the guide roll and one-third of the distance is from the guide roll to the lead-out fabric roll. In no case should the lead-in or lead-out draws be less than one-third of the face width of the fabric-carrying roll. Guiding can appear to be confusing, particularly with different degrees of wrap. One simple rule holds true under all conditions: A fabric will travel toward the end of the roll it contacts first.

A well-designed and efficient press should have a fabric guided with a straight tradeline. A tradeline that is not straight will produce distortion in the weave pattern which will result in nonuniformity across the width. This can negatively affect performance.

A tradeline that tends to run ahead at one edge is an indication that the fabric run is shorter at that end. This is corrected by tightening the stretch roll at that end to equalize the length of run. By doing so, the fabric may tend to run off the machine. The machine tender must watch this tendency and, if necessary, must correct it by repositioning the hand guide.

If the fabric wraps a concave or convex (crowned) fabric roll, the center of the fabric will tend to lag or to run ahead, respectively. This is caused by the difference in circumferential speed of the roll's center in respect to its ends. A fabric passing straight through the nip of two press rolls properly crowned for uniform nip contact without wrapping either of the rolls will maintain a straight tradeline. If, however, the crowns are excessive for the applied load so that the nip contact is greater at the midpoint of the roll face, the fabric will lead at its center. If the crowns are insufficient for the applied load so that nip contact is greater at the roll ends, the fabric will lag at its center.

If the fabric wraps one of the press rolls and the roll is crowned, the arc of fabric wrap will be greater at midface than at the ends. The rotational speed is the same across the face of the roll, so the fabric will run ahead at its center. The amount the fabric will advance at its center is proportional to the angle of wrap; this amount increases with increased wrap.

The tradeline position is generally the net effect of more than one cause of distortion. Knowing the above causes enables the papermaker to take corrective action by adjusting rolls to either remove the cause or to counteract it. Occasionally, fabrics from different suppliers may react differently on a given position due to stiffness and resistance to distortion characteristics inherent in different designs.

F. TROUBLESHOOTING

To obtain optimum press performance, considerable attention should be given to proper fabric operation. Following is a brief listing of some common operating problems which may interfere with pressing efficiency, along with suggestions for correcting them.

Dimensional Stability

If the fabric is too narrow, wide, long, or short, the first question is whether it started that way or became that way after running. Check to see if the operating tension is correct, as this will affect fabric size. Check records to see if previous fabrics exhibited a similar problem. See if the operating temperature has changed (e.g., steam box on). Check bowed roll applications. Examine the fabric for excessive wear. Look at tradeline distortion that may be present and make roll adjustments as needed.

Blowing

Blowing is caused by an accumulation of air between the sheet and fabric ahead of the press nip, and is greatly influenced by press section geometry. It can occur because air is entrained by the fabric surface, or because the fabric is too dense to permit air to evacuate, or because the felt is too open and air is being pumped through it. If a fabric blows when it is new but stops after it is broken in, it may be necessary to decrease the volume of the fabric by increasing density. A blow roll that raises the sheet above the fabric is one solution when it is practical. A suction lead-in roll or a suction box on the inside of the fabric just ahead of the nip is sometimes used.

Increasing the operating tension of a new fabric will sometimes alleviate blowing. Reduction in porosity as the fabric runs can cause blowing to develop. Consequently, cleaning of the fabric is helpful. If this condition persists regularly over the fabric's life, then a more open fabric construction is usually required.

Shankering and Shedding

If the needle batt fiber on the fabric surface is too long, it may tend to twist together and accumulate fiber stock. These accumulations are called pigtails or shankers and can result in sheet marking. The fabric maker can correct this by changing the needling procedure. As a temporary partial solution, the papermaker can stretch a rope across the fabric to drag over it to catch and remove some of the fiber accumulation. Occasionally, shedding will occur only on start-up due to stray, loose fibers. If shedding continues, it is due to improper needling or operating conditions on the paper machine. Some machine conditions to check include worn fabric suction box covers, lack of lubrication, high-pressure shower wear, chemical degradation (including high chlorine residual), and hydraulic damage in the nip.

Shadow Marking

When hydraulic forces are not uniform on the sheet during press dewatering, the so-called shadow effect of the pressure differential can be seen in the sheet. These marks can be from grooves in grooved rolls, from holes in suction or blind drilled rolls, and from the fabric structure. The use of a heavier or more rigid fabric is usually needed to adequately bridge the grooves or holes in order to provide uniform pressure. If the mark is from the fabric it is usually necessary to change designs. Other solutions include using softer roll covers, smaller holes, or narrower grooves, and in some cases recovering or regrinding a worn cover. Some of these solutions might reduce water handling capacity or sheet water removal. Also, filled fabrics and fabrics that are running too wet can contribute to shadow mark problems.

Marking-Yarn, Surface

When marking is more of a direct impression than a shadow effect, design changes are necessary. When base fabric yarns are marking, then yarn types, yarn density (count), and weave pattern may be changed to provide more uniform pressure to the sheet. Yarn marking can also be due to fabric wear. More paper side needle batt may be used to provide more coverage of the base. If the marking is from the fabric surface, finer fibers, different needling techniques, and precompaction treatments may be used to improve uniformity. Yarn marking can also occur when there are crown problems causing higher pressure in certain areas across the width.

For both shadow marking and fabric marking, it is necessary to determine which side of the sheet is being marked and the pattern of the mark. Knowing this can enable the papermaker to locate the cause. For example, closely spaced machine direction marks are likely associated with the fabric while wider spaced marks are typically groove roll marks.

Wrinkling, Roping

Taking care to operate the fabric at the proper tension across the entire width is the best way to avoid wrinkling and roping. Proper wet up procedures are particularly important to prevent uneven tension. On long unsupported draws, a supporting fabric carrying roll may be necessary. If wrinkling occurs, the press should be opened and the fabric spread out evenly. Steam or hot water may help to flatten the wrinkle. The press should then be lightly loaded and started up slowly.

Edge Flipping, Sheet Following

Problems with uniform sheet transfer can cause breaks and quality problems. The sheet tends to follow the smoother surface. Press geometry can also affect transfer, and sometimes changing the amount of wrap of the sheet or fabric on a roll will help. In some cases, a change of roll cover material is needed. The fabric may appear to be smoother because it has become filled, and in this case cleaning is effective. A wet running fabric surface will also create problems. Suction box and shower use can be modified to dry up the fabric. The fabric maker can modify the surface by using coarser batt fiber or by modifying the needling technique.

Sheet Pickup Problems

Poor pickup or drop-off may be caused by a number of problems including stock changes, condition of the forming fabric, condition of the press fabric, speed differential between the press fabric and forming fabric, incorrect suction pickup roll operation, and incorrect fabric conditioning equipment operation. Design of the pickup fabric is critical. The fabric must be open enough to allow vacuum in the suction roll to pull the sheet to it but it must have a closed surface that is smooth enough to hold the sheet to the transfer point. The surface characteristics are particularly critical in tissue applications. The moisture level in the fabric also influences sheet adhesion. If the pickup fabric appears filled or closed up, it should be cleaned. A plugged suction roll or improperly positioned deckles can cause problems, also. When problems are caused by poor edge cut by the trim squirts, running double squirts and adjusting the angle of cut is helpful.

Wear

The causes of excessive fabric wear can be divided into fabric design, mechanical, chemical, and hydraulic categories. A poorly designed fabric can experience wear even under normal operating conditions. Mechanical causes that should be investigated include:

- Press rolls - crowns, bearings, misalignment, rough or worn surfaces
- Fabric conditioning equipment - worn suction box covers, excessive shower pressures, oscillator operation, even coverage of lubricating showers, filled fabrics causing higher suction box vacuum and more drag
- Stock system - sand or abrasives in the furnish.

Wear in isolated streaks usually indicates a mechanical cause. When the wear is overall, chemical causes are suspect. These could include damage from chlorine residual in the system or from improperly applied fabric cleaning chemicals. Hydraulic damage can occur when the fabric/roll system cannot handle the existing water load. These hydraulic forces can literally blow the fabric apart. The press fabric, roll holes, and grooves should be checked for filling.

Crushing

The hydraulic forces discussed above can also disrupt the sheet. This crushing causes breaks and quality problems. On a new fabric, the problem may be that the fabric is too dense with insufficient void volume. A dirty, worn, or compacted fabric would have reduced void volume and could contribute to crushing. Isolated high pressure points increase the amount of water flow in that area. If the flow becomes excessive, crushing will occur. Crushing can also result from an excessive rate of pressure application within the sheet. Roll crowns and press loadings should be checked for uniform application of pressure. Filled streaks in the fabric or roll can also cause a problem.

Press Bounce

Press bounce and vibration can be caused by mechanical problems on the machine, nonuniformities in the fabric, and hydraulic forces. Checking the frequency is the first step in isolating the cause. Also, checking to see when the bounce started, whether it is getting better or worse, whether any rolls or fabrics have been changed recently, and whether the condition existed prior to these changes are all important questions to answer. Typical roll causes include improper balance, bad bearings, worn rolls, corrugated or flat areas in the cover, or insufficient void volume. Fabric causes are typically nonuniformities such as loom edges or batt splices. If the fabric is suspected to be the problem, deliberate cocking of the tradeline sometimes can temporarily alleviate the problem. Many mills regularly alternate the cocking. If the fabric or roll is filled or if the fabric design does not have adequate void volume, excessive hydraulic forces can cause bounce or vibration.

Wad Burning

Wad burning occurs when the sheet breaks and paper gets caught on the in-going side of the nip, where it spins and causes friction on the fabric. This friction creates heat which melts the fabric's surface fibers. The surface can also become plugged with fiber. Wool and some high-melting-point synthetics are more resistant to melt damage; however, these fibers do not offer the general wear and compaction resistance properties of polyamide. Press geometry must be changed or break detectors must be installed so that the press is quickly unloaded in order to eliminate wad burning. Poorly designed or misaligned roll doctors also contribute to wad burning problems. A wad burned area of the fabric sometimes can be cleaned by the use of chemicals and high-pressure showering. In extreme cases, a spiked wheel can be used to break up the fused surface. The wad burned area typically causes wet streaks in the paper because water cannot be pushed through the closed fabric surface.

Poor Water Removal

Press water removal that is less than normal can result in slow speeds, increased steam use, increased draws, crushing, marking problems, vibration, and increased breaks. Many of these problems have been discussed earlier. Analysis of a water removal problem can be very complicated due to a myriad of causes that can be stock, machine, or clothing related. Finding a solution starts with isolating the point when the water handling problems started along with any changes that had been made to the process or equipment around that time. It is important to determine if the problem is localized. The following areas should be checked: stock and wet end conditions, proper press loading, nip conditions (dry or wet, uniform or uneven), roll doctors and showers, all fabric conditioning equipment, and condition of the press fabrics. Moisture tests of the sheet at various points can help isolate the problem area.

G. SUMMARY

Whenever press fabric operating problems are encountered, it is recommended that you work closely with your suppliers. In addition to assisting in solving operating problems, suppliers may also observe conditions that necessitate a change in fabric design in order to improve performance.

Chapter 7

Felt Conditioning and Cleaning

Press felts seldom are removed because they are worn out. Replacement usually is needed when they are filled or compacted to the point where they can no longer handle water uniformly or they affect sheet quality. To maintain maximum press performance, all felts should be conditioned and cleaned continuously from startup to removal. Effective cleaning systems employ showers, and felt suction boxes as well as chemical programs.

A. CAUSES AND TYPES OF FELT FILLING

The press compresses the web and felt and transfers water and solids from the sheet into the felt, which leaves the press nip with more solids than when it entered. These solids have a varying affinity for the felt and must be removed or they can fill the voids in the felt structure. **Table 7.1** lists types of felt filling materials.

The nip compacts the felt with each pass. Press felts are very resilient, but they are not perfectly elastic and are subject to slow compaction over time. This also reduces the void volume, especially during the breaking-in period.

Inorganic Contaminants	Organic Contaminants
Silicates (clay, talc)	Pitch
Titanium	Hot melts
Calcium carbonate	Size chemicals
Iron oxides	Latex
Magnesium carbonates	Oil
	Defoamers
	Wet-strength resins

Table 7.1 Common felt contaminants

B. SHOWERS

There are four showers commonly employed for felt cleaning: high pressure, flooding, suction felt box lubrication, and chemical cleaning. These showers provide, respectively, mechanical action, flushing, lubrication, and chemicals to the felt. **Table 7.1** is a summary of felt shower use. Showers must be maintained for efficient press operation.

Shower water temperature and pH are important. The temperature should be maintained at or above the press operating temperature. The maximum temperature for synthetic felts is about 70° C (160° F), while that on the older type wool felts is 60° C (140° F). The pH should be maintained near the same pH as the machine headbox to avoid additional deposits or chemical shock, except for chemical felt washing during machine downtime.

SHOWER	PRESSURE (psi)	OSCILLATION NOZZLE	TYPE	USE
High pressure (Outside)	150 - 350 (1030-2410 kPa)	Yes	0.04 in. (1 mm) Jet or fan	Continuous
High pressure (Inside)	350 - 550 (2410-3790 kPa)	Yes	0.04 in. (1 mm) Jet or fan	Intermittent
Suction box	20 - 30 (140-210 kPa)	No*	90 - 120°F Fan	Continuous
Flooding	40 - 75 (280-520 kPa)	No*	30 - 60°F Fan	Continuous
Chemical conditioning	50 - 70 (345-480 kPa)	No*	Needle or fan	Continuous or intermittent

* Oscillated for streak-free cleaning

Table 7.2 Felt conditioning showers

High Pressure Oscillating Showers

High-pressure felt cleaning showers provide the hydraulic forces necessary to loosen the felt filling materials and open the felt surface by fluffing the batt fibers. These showers must be located ahead of any mechanical treatment such as felt suction boxes in order for loosened contaminants to then be removed. These showers are most effective if used continuously, with intermittent higher-pressure operation when the felt is highly filled. Shower pumps should be interlocked with the press drive so that the showers shut down when the felt is not running or when the oscillator fails.

Felts are compacted and soiled primarily in the batt layer, not in the base or mesh, so treatment must be on the face side. Thus, high pressure showers are usually located on the felt's face side and positioned near a felt carrying roll so that an outside felt roll does not come between the shower and the felt suction box. Face-side felt rolls force contaminants back into the felt. Needle type nozzles generally are used, but narrow fan-type showers can be used to reduce wear on the flap of seamed felts.

High-pressure showers are operated on a continuous basis at 200 - 300 psi (1370 - 2070 kPa) Intermittent higher-pressure (350 - 650 psi, 2410 - 4480 kPa) operation is useful for cleaning heavily contaminated felts. Continuous high-pressure operation may damage the felt batt.

High-pressure showers, especially needle showers, are oscillated to fully cover the felt and to avoid felt damage. The oscillation distance is in even increments of nozzle spacing for multinozzle systems, generally one or two times the nozzle spacing. Two times nozzle spacing is preferred because felt treatment still occurs if one nozzle plugs.

The frequency of oscillation for uniform coverage is related to the felt length and machine speed by the following formula developed at Westvaco (Equation 1):

$$R = \frac{S \times d}{L} \quad \text{and} \quad t = \frac{W}{R \times N} \qquad (1)$$

Where;

R: stroke rate for uniform coverage (in./min or cm/in.)
S: machine speed (ft/min or m/min)
L: felt length (ft or m)
d: shower stream width (in. or cm)
t: time for complete felt coverage (min)
W: felt width (in. or mm)
N: number of nozzles on shower pipe

The distance from the shower to the felt depends on the shower pressure and the shower nozzle diameter: The smaller the diameter, the closer the shower should be to the felt. At closer distances felt damage increases, and at longer distances the water jet is ineffective. A particular shower nozzle has an optimum cleaning range - about 4 - 8 in. (10 - 20 cm) from the felt.

The shower is angled 0 - 20° into or away from the felt run. Most current recommendations are to impinge against the oncoming felt, but impinging with the travel of the felt has also been successful.

Single-nozzle oscillating cleaning showers reduce fresh water consumption for felt cleaning and are gaining in popularity. Some have the additional feature of streak cleaning; that is, they may be programmed to operate on a narrow area of the felt that has plugged.

Inside high-pressure showers are used on tissue machine felts and other felts where felt marking and damage are important considerations. In general, inside showering cleans less effectively than outside showering. These showers are intermittently operated at 350 - 500 psi (2410 - 3450 kPa) pressure.

Flooding Showers

This type of shower floods the felt with water to flush contaminants from the inside of the felt to the outside. The current trend on modern machines is to do without the flooding shower and run the felts drier to avoid crushing in the nip. Flooding showers generally are used only on open or high-permeability felts.

Flooding showers are located inside the felt run, preferably to discharge into the nip formed between an inside felt roll and the felt. The showers are usually fan nozzles operating at 40 - 75 psi (275 - 520 kPa) with filtered white water. Traditionally these showers have been nonoscillated, but recently oscillators have been applied for streakless operation. These high-volume showers contribute largely to the felt water balance, so the pH temperature should be carefully monitored.

Heavy-weight felts limit the use of flooding showers because the volume of water needed to flood a heavy felt exceeds the capability of the felt suction box to remove water.

The maximum flow of the flooding shower is determined by the felt suction box capacity. The total shower water addition should not exceed the capacity of the felt boxes. A general rule of thumb is 0.06 - 0.10 lb (27 - 45 g) total shower water per lb (g) of felt. A better calculation would employ equations developed by DeCrosta to determine the maximum water removal for the existing boxes.

Suction Box Lubrication Shower

The shower is located immediately ahead of the suction box, on the face side of the felt, with the jet directed into the suction box-felt nip. Its functions are to lubricate the felt suction box, reduce the drag horsepower requirements of the press (*1*) seal the felt against the suction box cover, and evenly wet the felt. A fan shower operating at 20 - 30 psi (140 - 210 kPa) is usually applied. Recently these showers have been oscillated to ensure streakless felt operation. The nozzles should be spaced so as to lubricate the entire contact width of the box.

Additional lubrication showers often are located near the felt edge to prevent deckle edge wear and provide additional cleaning.

Chemical Cleaning Shower

The chemical cleaning shower should be located where the felt leaves the nip and well before the felt suction boxes. Both needle and fan showers are used. If needle showers are used, oscillator speed should be coordinated with machine speed for uniform application (see Equation 1). Chemical cleaning shower operation can be either batch or continuous mode, but the trend is toward continuous operation with increased white water closure.

C. MECHANICAL FELT CONDITIONING EQUIPMENT

Felt Suction Boxes

Felt suction boxes draw water and contaminants from the felt's face side. The water in the felt is pulled toward and into the slot by the high-velocity air flow through the felt. The felt boxes are generally preceded by mist or lubricating showers that lower the friction on the felt, evenly wet the cross direction of the felt, and add to the flushing action at the suction box. The water removed with the felt suction box is primarily the shower water applied to the felt during the cleaning process. Felt performance is indicated by monitoring vacuum level and air flow. Many mills use vacuum level to schedule felt replacement.

The most effective vacuum systems for felt suction boxes have independent pumps on each suction box or felt in the press section. This allows the vacuum level to meet the felt demand throughout its life. If a pump supplies two or more felts, air will take the path of least resistance and the more open felt will receive more conditioning than a more closed felt. An effective means of monitoring cleaning performance is to place independent drop legs and separators on each felt suction box, with calibrated weirs at the seal pit after the separator.

DeCrosta (*2,3*) developed empirical equations that can be used to determine vacuum and open area (Equations 2 and 3). When sufficient data are available, these equations are useful in determining the volume of air flow and vacuum level required throughout the life of the felt.

$$V' = \frac{0.069\ (dP)^{0.476}\ \ (t_d)^{0.110}\ \ (V^*)^{0.916}}{(M_{in})^{0.628}} \qquad (2)$$

$$M_{out} = \frac{1.23\ \ (M_{in})^{0.819}}{(V')^{0.024}\ \ (dP)^{0.124}\ \ (t_d)^{0.096}} \qquad (3)$$

Where;

V': specific air flow through the felt at the suction pipe (std. ft^3/min/in.^{2})),

dP: pressure drop across the felt at the suction pipe (in. Hg),

t_d: dwell time of the felt over the suction pipe (ms.),

V^*: felt permeability (ft^3/min/in^2 at 0.5 in. H_2O),

M_{in}: felt moisture ratio entering the suction pipe (lb water/lb felt), and

M_{out}: felt moisture ratio leaving the suction pipe (lb water/lb felt).

Felt suction boxes utilize three cover patterns: single slotted, double slotted, and herringbone. Much of the debate as to the merit of large single slots vs. double slots depends on the felt type in use.

Herringbone pattern covers can contribute to the felt spreading on less stiff felts, but not on newer, stiff felts. Herringbone covers require more air flow for the same conditioning but they are sometimes recommended for seamed felts to avoid excessive seam wear.

Most felt suction box covers are of HDP (high density polyethylene), or ceramic composition. The merits of each are listed in **Table 7.3**.

	HDP	CERAMIC
Capital cost	+	-
Cover life	-	+
Drag load	-	-
Ease of installation	+	-
Fabric wear	-	+
Thermal shock resistance	+	-

Table 7.3: Felt suction box covers

Multiple Shoe Type Felt Conditioner

The multiple shoe felt conditioner is a older felt conditioning method not commonly used in modern press felt conditioning systems. It consists of one or more small cleaning shoes mounted on a telescoping header and oscillated slowly across the width of the felt. Felt cleaning is accomplished by injecting water into the felt with a fine slot and removing the water with vacuum in subsequent slots. This system has the advantages of lower vacuum air flow requirements than full width felt suction boxes, and the ability to give preferential treatment to streaks. The disadvantages of the shoe system are its complex operation, the need for flexible hoses for vacuum and water, and the need for a reliable oscillator operating against the high vacuum drag resistance. This system is no longer manufactured, having been replaced primarily by the felt suction box.

Wringer Presses

Plain, grooved, or suction wringer presses used to be common methods of felt conditioning, but have been removed from most machines. The wringer is a pair of nipped rolls in the felt return run, generally preceded by a shower. The principle is that water and felt contaminants can be pressed out by a press operating only on the felt. The advantage of the wringer press is that the rotating press rolls do not drag on the batt surface, as does a felt suction box. The disadvantage is that the press adds to the felt wear by compacting it. The wringer press is often driven, adding to the complexity of the press operation.

Whipper Systems

Cylinder board machines have been equipped with a felt whipper positioned perpendicular to the felt travel on the felt face. The whipper rotates opposite to the direction of the felt travel, at a rotation speed three to four times the felt's linear speed. The whipper usually is preceded by a high pressure cleaning shower and followed by a flooding shower and felt suction box. The mechanical action of the high pressure shower and whipper sufficiently loosen contaminants so that the flooding shower can flush them from the felt. These systems are not common on new machines.

Fabric Brushes

Fabric brushes are sometimes used on the pickup felt to remove surface contamination, without causing severe mechanical disturbances to the batt layer. They consist of rotating brushes and attached showers that lightly scrape off contaminants. Showering removes the fiber from the brush. These brushes are most common on tissue applications where batt surface smoothness is of utmost importance.

D. CHEMICAL CLEANING

The use of chemicals to remove felt contaminants has two principle benefits. First, the void volume of the felt is maintained by removal of contaminants and is available for water removal. Second, many chemicals act as surfactants and decrease the flow resistance of the water in the felt, thus aiding press water removal and felt dewatering by the felt suction boxes. Too much chemical can be applied, thus disturbing the product performance, so the proper amount and types of chemicals are best worked out with the chemical and felt suppliers.

Various systems have been developed over the years to chemically clean felt. These systems have gained importance with the advent of higher press loads and drier operating felt conditioning systems. The four most common types of chemical conditioning methods are: off-machine in a washer, on-machine during shutdown, on-machine intermittent, and on-machine continuous.

The oldest and least used method is off-machine felt washing, which is limited to machines with special marking felts that are regularly taken off between orders. This method is the most effective because washing time can be optimized and a wide range of chemicals can be used without interfering with the machine operation.

On-machine shutdown cleaning is performed on most paper machine press felts. This method allows strong chemicals to be used for short periods of time, without contaminating the machine water system. The cleaning is usually done with full width showers while the machine is slowly driven or crawled. The choice of acid or caustic washes with detergents depends on the types of contaminants. Water used in this system should not be returned to the white water system because chemical shock often results. The disadvantage of this method is that felt condition deteriorates between machine shutdowns, so it often is used in conjunction with chemical application during operation.

With on-machine intermittent cleaning the chemical is applied on a batch basis, for example once a shift. The cleaning chemicals are applied through full width cleaning showers located as soon after the press nip as possible. Whenever chemical solutions are applied while the machine is running, care should be taken that these chemicals do not adversely affect the paper sheet or press operation. Although stronger solutions may be used with this method, the felt cleanliness, as indicated by felt permeability, fluctuates between cleanings.

Continuous chemical cleaning is rapidly becoming a standard for felt conditioning. Full width showers apply dilute chemicals to prevent buildup throughout the felt life. A device such as a metering pump injects chemicals at a constant rate into the shower's water supply. Acid and alkaline solutions are used to remove different types of contaminants.

The treatment should be selected carefully because chemicals added to the felt for cleaning are in contact with all press components as well as the paper web. Suppliers of felts, roll covers, and chemicals should be consulted about the possible adverse effects that may result from a particular treatment.

References

1. Neun, J. A., Tappi 64(9):123 (1981).

2. DeCrosta, E. F., Tappi 56(11):100 (1973).

3. DeCrosta, E. F., Preprints of 1975 International Water Removal Symposium, BPBIF, Technical Section, London, England.

Chapter 8

Instrumentation for the Press Section

A. INTRODUCTION

The use of instrumentation to monitor press section performance is a necessary feature for the efficient operation of a paper machine. The tremendous energy cost to thermally remove water in the dryer section has imposed a most urgent need to maximize mechanical water removal in the press. By monitoring and regulating the various processes and equipment variables which influence dewatering, the operator can optimize water removal while building the sheet's physical properties in a controlled fashion.

Process instrumentation for the press section includes those devices for measuring water in the sheet and felts, speeds and draws, nip loads, and felt tensions and conditions. A typical modern paper machine would include moisture, load, and speed/draw measurements as part of a computer process control package to monitor production parameters. This information might also be used to adjust certain machine parameters in order to minimize cost, maximize production, or automate grade changes.

The ultimate method for measuring press performance is to sample the sheet before and after the press. This is rarely used on production machines because of safety concerns and downtime constraints. This technique is addressed at the end of this chapter. (See Section on Sheet Sampling.)

B. MOISTURE CONTENT AND TOTAL MASS

The measurement of sheet and felt moisture content is useful for determining dewatering performance, as well as for developing material balances around the machine. Scanning moisture gauges are valuable devices for monitoring and correcting cross-direction (CD) profiles, in addition to identifying machine-direction (MD) disturbances. Moisture gauges typically utilize electromagnetic waves in infrared, microwave, or radio frequency techniques. Frequency selection depends upon sheet basis weight and moisture ranges. There are also differences between commercial products. Signals from moisture gauges are usually related to the amount of water per unit area. To convert this number into sheet percentage moisture, the basis weight must also be known.

Basis weight gauges, occasionally used in the press section and almost always at the dry end, use both infrared and nuclear radiation transmission techniques.

Table 8.1 shows a summary of the types of radiation and wavelengths used in the various applications.

Infrared (IR) Moisture Gauges

Most infrared moisture gauges measure moisture by comparing the absorption of two radiation beams that are only slightly different in wavelength. One wavelength (the reference) is insensitive to water, whereas the other (the measure) is highly absorbed due to vibrational resonance of the water molecule. The ratio of reference to measure signals is proportional to the weight of water present in the signal path. Both single-sided and two-sided gauges are in use.

A typical single-sided, infrared gauge arrangement, mounted in a single head package, is shown in **Fig. 8.1.** A quartz-halogen source is focused on the sheet, reflected back, split, filtered, and collected with the detectors. Some manufacturers use a single detector. This reflected, or scattered-mode, technique measures total sheet moisture for light papers, such as the printing grades. For heavier grades only the surface moisture is measured, which due to moisture stratification in the sheet may not be representative of average sheet moisture.

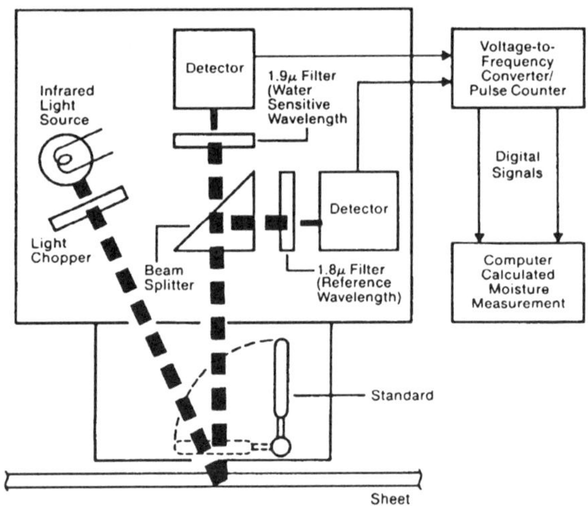

Fig. 8.1: Single-sided infrared moisture gauge
(Courtesy Measurex Corp.)

A typical gauge sees a spot of about 20 mm diam., which is sufficiently small to detect moisture streaks. Gauge compensation and standardization eliminate measurement error from such sources as temperature, dust buildup, and passline variations (sheet distance from the gauge).

One unique feature of some of these gauges is the determination of percentage moisture directly without the need for a basis weight gauge. Since the typical weight gauge uses nuclear radiation, it is affected differently by refining, pressing, and calendering than an infrared gauge. The use of infrared energy for combined moisture and basis weight eliminates the need to compensate for these effects on the different types of gauges. This combination IR gauge, however, depends on the sheet scattering coefficient staying constant, at the value for which the gauge was calibrated.

Two-sided or transmission IR gauges are used in the lighter basis-weight ranges (<200 g/m^2), especially in the recycled and dyed grades. As shown in **Fig. 8.2**, instead of capturing back-scattered energy the gauge has two hemispherical cavities in which IR energy is continually reflected until detected. Here, the proportion of the water-sensitive wavelength's energy which passes through the wet paper is inversely related to the amount of water present per unit area. Multiple sheet passes are used to raise sensitivity. The operating principle of this gauge is otherwise identical to its single-sided cousin, with the exception that average sheet moisture is necessarily measured because the entire sheet thickness passes the IR energy.

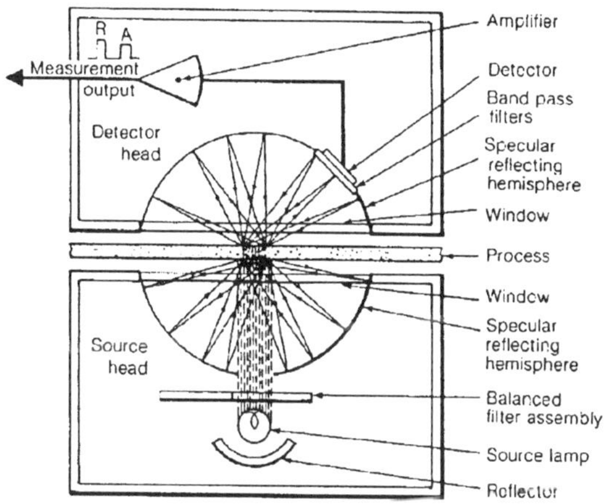

Fig. 8.2: Transmission infrared moisture gauge
(Courtesy ABB, Inc.)

Microwave Moisture Gauges

Most microwave gauges, used in the heavier basis-weight grades (>200 g/m^2), use the principle of cavity resonance which is influenced by the presence of water molecules and their dipole rotation. These gauges measure resonance frequencies, where the difference in two measurements is proportional to the amount of water per unit area in the viewing field of the instrument. These gauges are available in either a single-sided contacting or a two-sided noncontacting variety. They can be used for either sheet or felt moisture measurements; a separate weight measurement is needed for determining percent moisture. Some microwave gauges use a microwave absorption principle where the amount of microwave energy transmitted to a detector is measured.

A single-sided resonance-type gauge is shown in **Fig. 8.3.** The gauge broadcasts at a certain reference microwave frequency (about 2 GHz) at which the water molecule has no effect. The gauge then broadcasts at a range of frequencies for which the rotational mode of the water molecule will respond. The presence of water effectively changes the dielectric constant of the system resulting in a different resonant frequency. The frequency shift of resonance from the reference is thus a function of the quantity of water present. The single-sided contacting gauge is losing favor because of potential buildup problems, and the fact that the gauge is sensitive to moisture stratification and sheet surface characteristics. Still, it is the most widely used gauge on press fabrics.

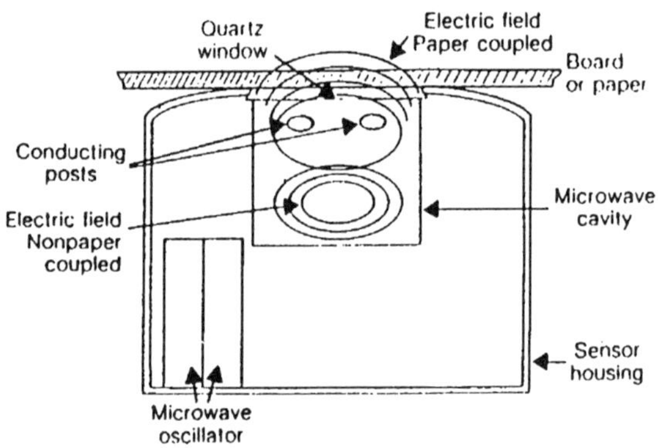

Fig. 8.3: Single-sided microwave moisture gauge
(Courtesy ABB, Inc.)

A double-sided gauge could be thought of as two noncontacting single-sided gauges. Actually, the gauge consists of a microwave source, receiver, and half of a resonant cavity on one side of the sheet, and a passive half of a resonant cavity on the opposite side. The resonance frequency shift is thus a function of the average sheet moisture across the full thickness of the sheet. This type of gauge is relatively insensitive to moisture stratification and sheet z-direction density variations, but exhibits sensitivity to head cavity separation distance.

Radio Frequency (RF) Moisture Gauges

These gauges contact the sheet, establishing alternating fields in the sheet as it passes over the gauge's cross-machine electrodes. Due to a combination of charge transport (interfacial polarization) and the water molecule's dipole orientation because of the radio frequency radiation, the admittance of the system changes with the amount of water per unit area present. A RF current detector provides the signal.

Compensation for sheet temperature is required and is provided by a separate infrared sensor. These gauges are especially useful at low moisture levels with the heavier paper grades where microwave gauges lose sensitivity.

Basis Weight (BW) Gauges

Basis weight gauges measure the total mass of material in the viewing path of the gauge. As shown in **Fig. 8.4**, a typical gauge system will consist of a focused source of beta radiation on one side of the sheet, an air gap, and an ion chamber detector on the other. As the emitted radiation is absorbed by any material in the path to the detector, changes in basis weight are measured as changes in detector signal. A typical measurement spot of 25 mm diam. allows good streak detection.

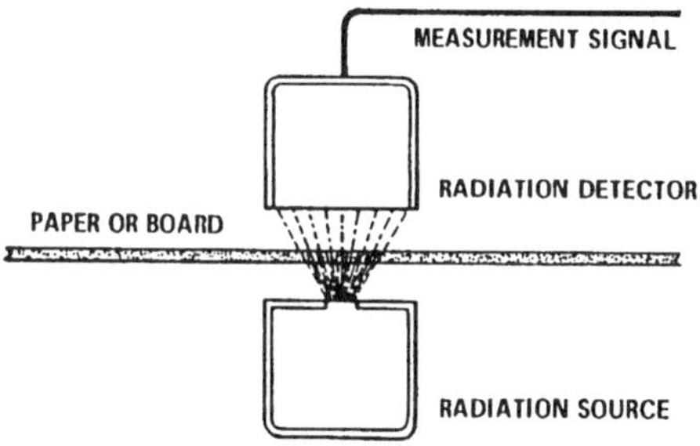

Fig. 8.4: Basis weight gauge
(Courtesy ABB, Inc.)

Different radiation sources, air gaps, and ion chambers are used as a function of sheet basis weight. Most grades depend on beta-ray absorption caused by ionization interactions in the outer electrons of an atom. Higher-energy gamma radiation typically interacts via the photoelectric effect in bound inner electrons, or causes displacement of nucleons in an atomic nucleus. The energy level of this radiation is many orders of magnitude greater than that of infrared because of the need to penetrate the sheet. Infrared can be used in the tissue grades, with a nucleonic source of X-ray photons used for the very heavy grades.

C. SPEED AND DRAW

Speed measurement of rotating equipment has been made much more accurate and reliable in just the past few years because of advances made in optics and electronics. The need for accurate speed measurement and control is demanded as machine speeds increase. Also, precise regulation of each section of the machine is necessary for good sheet transfer and property development.

Electromechanical Devices

These devices all use the familiar principle of a moving magnetic field inducing a voltage in a pickup coil placed near the circumferential path of the magnet, as shown in **Fig. 8.5**. Assuming the magnet is attached to a paper machine roll, the strength and frequency of the voltage is a function of the speed of the roll.

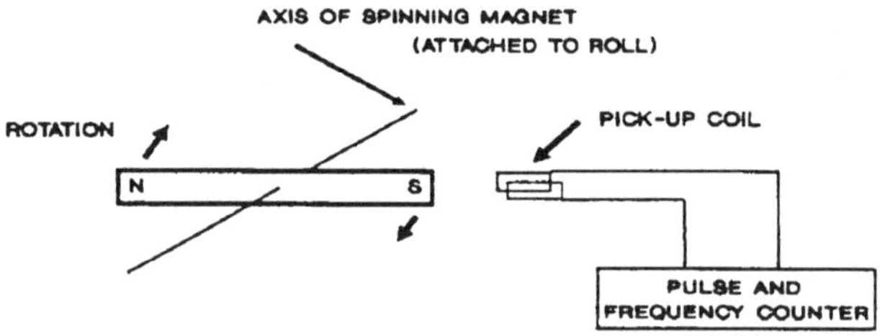

Fig. 8.5: Speed measurement

Typically, a pulse or frequency counter would be used to convert pulse frequency to roll speed. Resolution of this system is increased by increasing the number of magnetic poles and pickup coils. The tachometer-generator system is one of a multitude of these devices.

Another category of device is the Hall-effect gauge, which is in fact a proximity sensor. Placed near the gear teeth of a driven roll, this sensor provides a pulsating voltage that detects passing teeth. Speed information is provided by a frequency counter.

Optical Encoders

These devices afford the user a more mechanically robust and precise speed measurement tool. Here a disk is spun by attachment to a paper machine roll. This disk has 1000 to 2000 scribed lines which interrupt the reflection of light emitted by a light-emitting-diode (LED) and detected by a phototransistor. The number of pulses in a time interval is counted and converted to frequency, roll speed of rotation, and roll surface speed. Upwards of 50,000 pulses per disk revolution is achievable with external electronics, resulting in speed resolutions in fractions of meters per minute.

D. NIP LOAD AND FELT TENSION

Nip load is an important process variable which affects both dewatering performance and physical property development. The advoidance of crushing is important for all grades, while the attainment of the highest dryness is an economic necessity. Felt tension is important for proper pickup, pressing performance, and felt life. Most systems for load measurement, whether nip load or felt tension, either measure the load directly with a load cell, or use a calculation scheme for load based on an applied pressure measurement.

Applied Pressure Schemes

The load that is applied to a roll, whether nipped or carrying a felt, is usually applied via a hydraulic or pneumatic cylinder arrangement. The pressure in the cylinder can be converted to nip load or felt tension by calculations which consider the mechanical configuration of the system.

The problem thus resolves itself to the measurement of fluid pressure. Typical pressure sensors arrangements are shown in **Figs. 8.6 to 8.10.** Principles of measurement may be based on the technologies of strain gauges, diaphragms and springs, liner displacement measurements, or optical techniques.

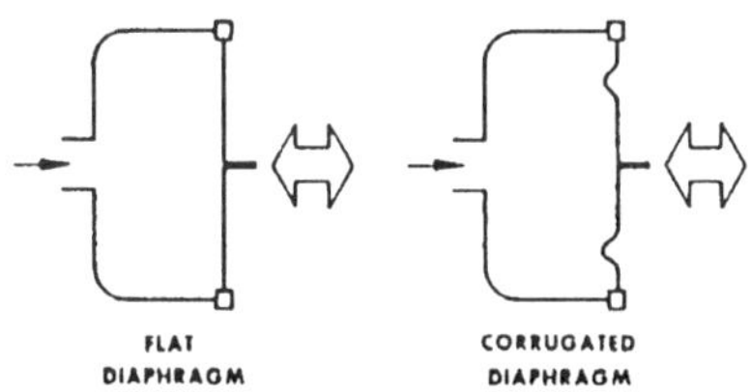

Fig. 8.6: Diaphragm pressure sensors
(Courtesy Rosemont, Inc.)

Strain gauges are devices bonded to the deflecting disk (or diaphragm) where the strain produced by the fluid pressure changes the electrical resistance of the gauge. The strain gauge is placed in a Wheatstone Bridge for measurement purposes (see **Fig. 8.7**).

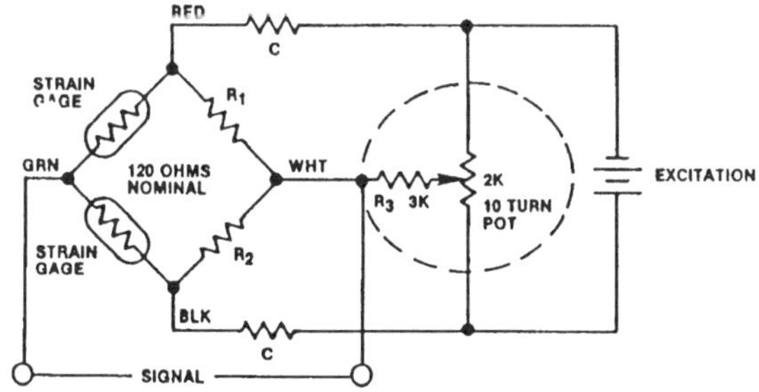

Fig. 8.7: Wheatstone bridge circuits
(Courtesy Data Instruments, Inc.)

Capacitance gauges provide a variable capacitance as the diaphragm moves (see **Fig. 8.8**). This change is capacitance is sensed by oscillator circuitry to produce a pressure measurement.

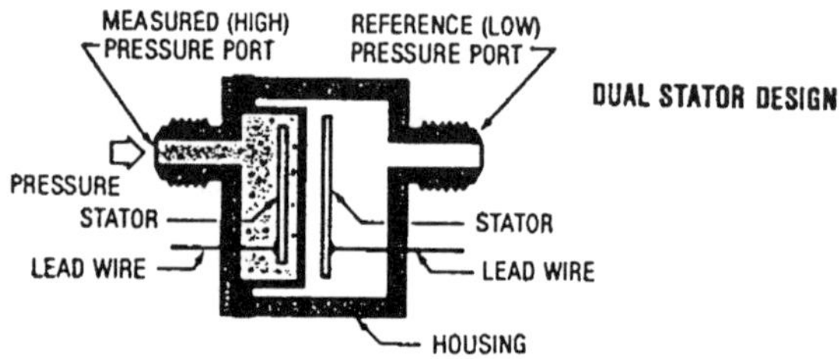

Fig. 8.8: Capacitance pressure gauge
(Courtesy Transamerica De Laval, Inc.)

Linear variable differential transformers (LVDT's) sense disk movement by attaching it to the core of a transformer. This device can also be attached to a familiar Bourdon tube, (see **Fig. 8.9**). As the core moves from the center position, the voltage induced into the windings changes, providing a voltage signal proportional to movement.

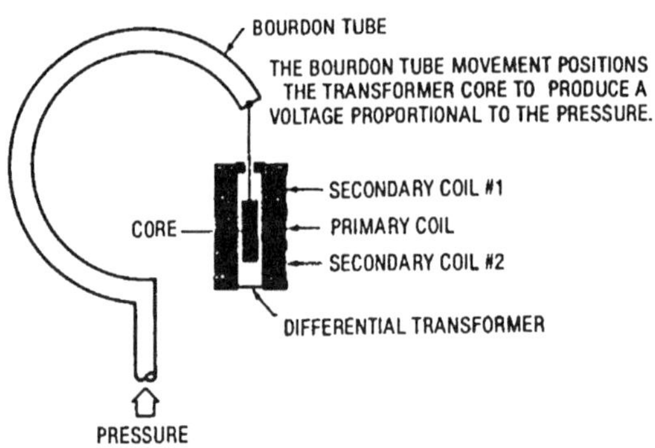

Fig. 8.9: Bourdon tube/LVDT pressure gauge
(Courtesy Transamerica De Laval, Inc.)

Optical pressure sensors are configured as shown in **Fig. 8.10**. A source of light, usually a LED, is positioned across from a phototransistor. As the pressure changes the diaphragm deflects, and the amount of light captured by the phototransistor changes.

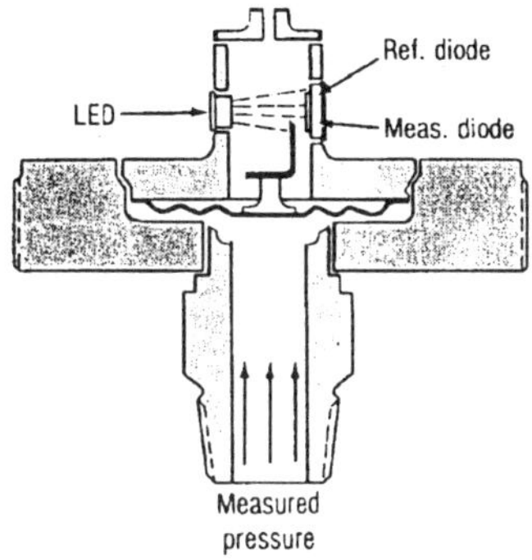

Fig. 8.10: Optical pressure sensor
(Courtesy Dresser, Inc.)

Load Cells

In many ways a load cell is a pressure sensor where the deflecting disk is backed by a stiff spring. As a load is placed on the disk, the disk deflects an amount proportional to the applied load. Thus, all of the methods for detecting disk movement for pressure transducers apply equally well to load cells. Instead of a load cell *per se*, press load is sometimes measured by mounting a strain gauge on the bearing pedestals.

Felt Tensions

Average felt tension is measured as a load on a partially wrapped roll by any of the methods discussed above. Cross-direction profiles of local felt tensions can be measured during operation with a hand-held portable device consisting of a spring-loaded moveable shoe that is pressed into the felt. The deflection of the shoe, read from a dial indicator, provides the tension measurement.

In many cases, the actual stretched length of the felt is considered a measurement of felt tension. The length is indicated by the position of the felt stretch roll. This measurement can be obtained by attaching the wipe of the potentiometer to the stretcher roll to provide a variable electrical resistance proportional to stretch roll position. This measurement must be viewed with suspicion, particularly during the initial running of a new felt when its length changes considerably.

E. NIP PROFILES

Actual nip profiles of presses while operating cannot be obtained, short of mounting pressure transducers across the machine in a press roll. Good clues as to profiles are obtained from scanning basis weight and moisture gauges, as well as variable crown roll settings and felt behavior. Temperature measurement via infrared pyrometry of the felt or roll surface can also be used to indicate profile problems. These measurements can have high errors due to the presence of water vapor.

Static nip profiles are taken by loading the nip with impression paper (both with and without the felt), which gives a visual indication of nip uniformity and width.

A recent development for this technique utilizes press film consisting of colored dye capsules imbedded in the film which rupture at various unit pressures. Here actual static loads and profiles can be obtained by using a color meter to measure the film. Hydrostatic controlled-crown rolls can be rotated while the film is nipped to provide information about barring.

F. FELT FILLING AND VACUUM

The degree of felt openness, or permeability, is an important factor in the hydraulic flow resistance in the press nip. As the porosity of the felt changes during operation, sheet transfer and dewatering performance are affected, resulting in changing operation on the machine. Measuring felt permeability on the run is at best a technique that yields only qualitative results.

Uhle box vacuum can be used as an indicator of overall felt filling if the box is supplied by one vacuum pump. As the felt fills, the vacuum rises due to lower air flow. Good seal is a necessity, and confusing results can be obtained if several vacuum elements are supplied by the same vacuum pump.

Several hand-held or traversing devices can be used to obtain CD felt filling profiles. A small traversing Uhle box can be used to obtain information about felt permeability at various positions across the machine. Similarly, an air shoe pressed against the felt can be used to relate compressed air flow to felt permeability at various positions across the machine. Permeability CD profiles are very important tools in the analysis of wet streaks or similar nonuniformities.

Vacuum gauges and transducers tend to be either the Bourdon-tube type for visual indication, or a deflecting-disk pressure gauge calibrated for the vacuum regime.

G. SHEET SAMPLING

Sampling the sheet before and after the press nip is, of course, a definitive way of obtaining press performance. Still, this technique is not without its own set of errors, to say nothing of the safety concerns when sampling. The method also suffers from the inability to sample anything but the extreme edges of the sheet.

One method for sampling involves using an air lance on the back side of the conveying felt to blow the sheet into a suitable receptacle. Questions raised about this procedure include whether any dewatering is done by the air knife, or whether water is blown into the sheet from the felt.

Another method "doctors" the sheet off the felt with a convex piece of plastic held against the sheet side of the felt. The sheet streamer is then captured and analyzed for moisture. The question again is whether the doctoring action dewaters the sheet or preserves its true moisture.

These methods frequently cannot be used in compact, multiple presses that do not allow reasonable access. This is an area where pilot machine sections can be used to obtain performance information as operating variables are adjusted.

Table 8.1: Types or Radiation and Application Features of Gauges

Type of Radiation (per unit area)	Wave Length (m)	Frequency (Hz)	Process Variable	Measured Variable	Main Application
Infrared (Scatter mode)	1.8×10^{-6} and 1.95×10^{-6}	10^{14}	Water	Signal strength ratio between the two reflected wavelengths	Single-sided surface moisture and weight
Infrared (Transmission)	Same as above	10^{14}	Water	Signal strength ratio between two transmitted wavelengths	Double-sided moisture for lighter grades and lower moistures
Microwave (Resonance)	10^{-1}	3×10^{9}	Water	Frequency shift of new resonance compared to reference	Single- and double-sided moisture for heavier grades
Microwave (Absorption)	10^{-2}	22×10^{9}	Water	Signal strength at detector	Double-sided moisture
Radio Frequency	10^{3}	3×10^{5}	Water	Current with RF detector	Single-sided moisture for heavier grades and lower moistures
Beta Rays	5×10^{-10}	10^{18}	Total mass	Current due to gas conductivity in ion chamber	Double-sided basis weight for most grades
X Rays	5×10^{-11}	10^{19}	Total mass	Same as above	Same as above for very heavy grades

Chapter 9

Auxiliary Press Equipment

A. DOCTORS

General

Doctor blades and/or wipes are used in the press section to perform one or a combination of the following functions: profile control, cleanliness, or aid in dewatering through control of rewetting. These functions aid in maintaining good machine runnability and sheet quality. Depending on the specific function or functions of the doctor or wipe, its positioning, material, configuration, and running conditions will be determined based on its compatibility against the roll on which it will be used.

Press Roll Doctors

All press rolls should be doctored. The rolls in contact with the sheet are doctored to prevent the roll from being wrapped with the sheet when the sheet is being threaded through the press or during a sheet break. It also keeps the roll surface free of crumbs that may be picked from the sheet. On suction press rolls and grooved rolls, doctors are now being used to eliminate rewet in the in-going nip from water following the periphery of the roll. These doctors should be oscillated slowly across the width of the roll face to minimize localized wear, prevent scoring of the roll surface, and prevent crumbs from passing under the doctor blade. If the sheet doctored off the roll cannot fall directly into a press pit, the doctor is usually fitted with a large apron to catch and hold the broke as it accumulates. The broke is then removed by the machine tender or one of the machine hands. Appropriate arrangements should be made to handle the broke in situations where there is no pit under the doctor blade on the first press roll, as with a belt conveyer.

Press Fabric Roll Doctors

Doctors are placed on some press fabric rolls for various reasons. A doctor on the first outside press fabric roll after the press nip can doctor off any part of the sheet that may follow the press fabric when passing through or during a sheet break. Doctors may also be placed on the inside and outside press fabric rolls. This will doctor off any fibers and/or fillers that may bleed through the press fabric or that are picked off the press fabric and deposited on these rolls. This, in turn, maintains machine cleanliness and reduces the rate at which a press fabric may fill up and thus inhibit its dewatering capacity.

Suction Roll Wipes/Deflectors

Although a wipe is not in a pure sense a doctor blade, it fulfills some of the same functions. The suction roll with holes and/or grooves can have high volumes of water following its periphery, causing potential rewetting of the press fabric and sheet in the in-going nip. A wipe, usually of rubber, contacts the roll surface; it is positioned ahead of the point where the press fabric contacts the suction roll and wipes or deflects this surface water into a saveall pan. A wipe is usually attached to the press saveall pan. A deflector works in the same fashion except it does not contact the roll; it is placed less than an inch from it. A purge shower is sometimes used to remove the rimming water.

B. PURGE SHOWERS

Fabric Purge

In a fabric press, water is transferred from the sheet and press fabric to the fabric voids. This water is then discharged via negative pressure drop at the exiting side of the nip, by centrifugal forces of the fabric wrapping a fabric return roll, and by use of a suction pipe or a purge shower. The purge shower uses compressed air, and the nozzle spacings and pressure are based on fabric design and machine speed. However, this method is not as cost-effective as using a suction pipe.

Fabric Shrink Sleeve Purge

Water expressed from the sheet and press fabric into the voids of the shrink sleeve fabric on a roll can be removed by means of an air purge shower. Closely spaced jets of compressed air impinge on the fabric roll surface at an appropriate angle, usually 10° to 30°, against the direction of the roll's rotation.

Grooved Roll Purge

A grooved press roll entrains water in the grooves that has been expelled in the nip. In order to remove the water from these grooves, a water purge shower with fan nozzles is directed against the roll's surface. The shower pressure, angle of impingement, and distance from the roll are determined according to speeds and water volume in the grooves. The angle of impingement, usually 25 to 45°, is directed against the roll's rotation. A wipe and/or doctor may be utilized after the shower to direct the water into a catch pan.

C. INTERNAL SUCTION ROLL SHOWERS

On many machines where insufficient water is expressed by the nip into the roll's suction box, it is necessary to lubricate the internal shell surface with a water spray. The volume of water spray should be kept minimal to prevent rewetting of the press fabric in the in-going nip. Self-lubricating seal strips are now available, thereby eliminating the need for internal showering.

D. PRESS FABRIC CLEANING SHOWERS

Press fabric cleaning showers are located in the press fabric return run for continuous and/or intermittent cleaning and conditioning. The showers may be high-pressure fan, needled nozzles, or chemical-fan nozzle types. The use of high-pressure showers is covered in detail in Chapter 7. The chemical cleaning shower may include application of detergent, acid, or alkali to clean and maintain press fabric openness. The type of detergent or the concentration of acid and/or alkali used depends upon the type of filling to be removed and the press fabric fiber content. Rinsing water should be acidified to avoid plating out of filling in the press fabric. Press fabric cleaning also is covered in Chapter 7.

E. SLOTTED BLOW BOXES

Some press arrangements utilize a slotted blow box over which the press fabric and sheet slide (Fig. 9.1). This equipment removes any boundary layer of air ahead of the nip and thus prevents blow wrinkles in the sheet. Suction press fabric rolls may also be used for this purpose to ensure a uniform sheet transfer in open draws in the press section. The effectiveness of these devices at reducing blowing diminishes as a press fabric fills up.

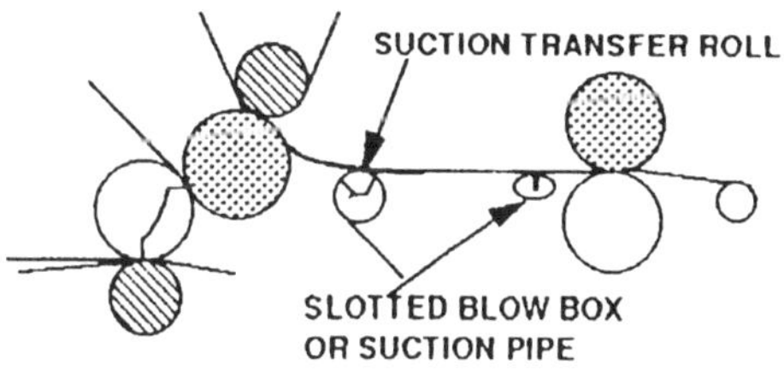

Fig. 9.1: Blow box on fourth press

F. SUCTION PIPES

Suction pipes, commonly referred to as uhle boxes, are utilized in the press fabric return run. Their function is to dewater the press fabric of water entrained from pressing action by the nip and/or added by lubrication and high-pressure showers.

Suction pipes commonly are located on the sheet side of the press fabric in the return run closest to the exiting nip, where practical. The pipe cover material varies from polyethylene to high density polyethylene to ceramics. The covers may be in the form of blades, rods, or herringbone construction. The appropriate dwell time, open area and applied vacuum required for dewatering must be established. These variables depend on the press position, volume of conditioning water, and volume of water from pressing. Dwell time is dependent on machine speed and suction pipe open area, which may be established via multiple pipes, multiple openings per pipe, and the size of each opening. The suction pipe should be designed so it may be turned away from the press fabric; this will enable the cover to be cleaned of any pitch or contaminants that may build up. Vacuum typically applied ranges from 5 in. to 20 in. Hg (mercury), depending on the application.

G. WRINGER PRESSES

A wringer press is located in the press fabric return run as a means of removing water from the press fabric, and for conditioning. It is largely being replaced by suction pipes. The wringer press is on the decline because it causes reduced press fabric life, greater rate of caliper reduction, and limited use of press fabric designs due to compressibility requirements. Also, it requires much maintenance, and it can cause sheet profile problems. Wringer presses include two mating rolls of various types. The press fabric is run through the nip of the two rolls, which dewaters it. The amount of dewatering is dependent on nip pressures and any vacuum applied through the rolls.

H. STEAM SHOWERS

Steam showers are being utilized in the press section to improve sheet dewatering and sheet profiling. Generally, the steam shower or box is located so that it is directed against the sheet and press fabric as it wraps a press roll, or against the sheet as it wraps a solid roll. Careful operation is important to ensure safe and effective results. Consideration of roll types, roll vacuums, steam shower temperatures, steam box positioning, steam box width, and steam pressures must be taken into account for proper operation. The steam shower may have various compartments through which steam temperature and pressure for profile control can be regulated.

I. WATER REMOVAL MONITORING

In the press section, dewatering occurs in the press nip and/or suction pipes. Dewatering is dependent on speeds, nip loadings, roll and suction pipe vacuums, suction plate, open area, dwell times, and press fabric designs and ages. To understand the interrelationships among these variables, it is crucial that the volume of water removed at a press nip vs. a suction pipe is determined. It is also of interest to determine when the effective dewatering capacity of a press fabric has ceased due to age or filling.

The most practical and inexpensive method of monitoring suction pipe and roll saveall pan flows is through the use of a weir (**Fig. 9.2**). The weir must be properly sized to handle the expected water flow range of what is to be measured. Collecting weir measurements twice weekly will yield sufficient data to trend press fabric compaction, filling, inappropriate design, or a change in machine operation. For suction pipes, the weir is located in the piping after the separator, if one is present.

Another flow-monitoring method uses ultrasonic gauges (see **Fig. 9.2**). These are more expensive than weirs and require a properly designed vacuum piping system to ensure a full flow in the pipe. A gauge is usually strapped on piping exiting the suction pipe separator pump.

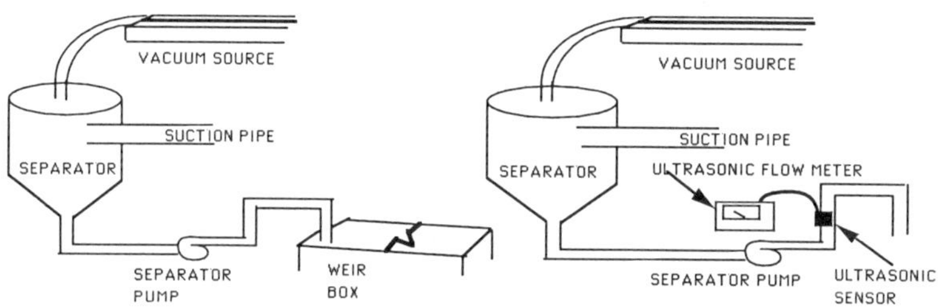

Fig. 9.2: Monitoring suction pipe flow via weir box or ultrasonic flow meter

Chapter 10

Factors Influencing Press Water Removal Performance

Water is removed from the sheet by both the compressive action of the nip formed by two contacting rolls (or a roll and a mating shoe) and the compressive response of the sheet under transverse load. The amount of water removed is a complex function of a large number of factors. Among them are felt characteristics, roll covers and venting, rewet, in-going sheet moisture, basis weight, temperature, press impulse including pressure uniformity, and furnish properties.

A list of all the possible factors involved in the stable operation of the wet press on a paper machine is exceedingly long. The important factors in press operation are many and include avoidance of sheet crushing; influence on sheet quality; roll crowns and cross-machine moisture profiles; prevention of press vibration; and others that are covered elsewhere in this manual. This chapter is concerned with those variables most directly related to water removal performance in the press operating under steady-state conditions.

A. PRESS CHARACTERISTICS

Several authors have compiled lists of factors believed to be involved in water removal. Robertson *(1)* presented a diagram in 1976 of 10 variables. This was expanded to include 20 variables and was published by Bliesner *(2)* in 1978 and later by Cox and Robertson *(3)*. In 1981 DeCrosta *(4)* compiled a diagram that included nearly 40 possible variables. The lengths of these lists illustrate the need to identify the relative importance of each variable in water removal so that both current application and future research efforts will concentrate on the most responsive factors.

In 1969 Wahlstrom *(5)* introduced his wet pressing equations, which recognized flow resistance in both the paper and felt, the problem of uneven pressure distribution, the role of sheet rewetting, and basis weight as major factors. In Wahlstrom's analysis, flow of water out of the fiber was considered a minor term. In the latest model *(6)*, he recognizes the importance of a flow resistance component in the cell wall as critical to effective water removal. This later analysis includes the role of felt rewetting, and separates the dynamics of sheet rewet into expanding-nip and post-nip portions. Bullerwell, Sutherland, and Krotchko *(7)* list as primary variables the nip dwell time, nip pressure, temperature, felt and roll cover design, and sheet properties. Uniformity of pressure, flow resistance, sheet basis weight, water retention value, and rewetting are mentioned as additional factors in press water removal.

This diversity in the identification of primary variables and an ever-growing number of them underscore the importance of efforts to identify the relative importance of the principal variables governing water removal in the press. The purpose of this chapter is to sort the primary and secondary factors that influence wet press water removal from the long list of proposed candidates. However, prioritizing variables is risky as any classification will be based on inadequate experimental data. In addition, there always seem to be data that are purported to be the exception to the rule. This present attempt to evaluate the relative importance of various wet press water removal factors is based on our current technical understanding. A summary of variables is given in **Table 10.1**.

Table 10.1 Wet press variables in water removal

Primary	Secondary
Post nip rewet	Initial felt flow resistance
Ingoing sheet dryness	Ingoing felt dryness
Grammage	Rewet in the nip at speed
Furnish properties	Shape of pressure profile
Double felting	Roll cover
Sheet temperature	Felt design
Impulse (Pressure x time)	Roll venting
Felt pressure uniformity	

B. PRIMARY VARIABLES

A primary variable is defined here as one capable of influencing outgoing sheet dryness by 4% or more where large changes in the variable can be achieved. State-of-the-art wet presses have been designed to apply principal influence over the primary factors involved in water removal. The extent to which the press designer or the machine operator can exert influence over a press variable will be limited by the behavior of the materials in the press nip (e.g., sheet, felt, and press roll performance), and by the cost-effectiveness of the design or operating condition.

Press Loading

Nip load (pounds per lineal inch, pli, or newtons per meter, N/m) is the first of several primary variables. Though higher nip loading decreases felt life and increases the press drive load, these losses are offset by appreciable gains in moisture removal and improved sheet consolidation leading to the development of better, more-consistent sheet strength.

Press loading is commonly expressed as the loading per unit cross-machine length of nip, i.e., kilonewtons per meter (pli) of roll face, as in equation (1).

$$\text{Nip Load (kN/m)} = \frac{\text{Total Roll Load (kN)}}{\text{Width of Roll Face (m)}}$$

or (1)

$$\text{Nip Load (pli)} = \frac{\text{Total Roll Load (lb)}}{\text{Width of Roll Face (in.)}}$$

However, press action is a function of nip loading per unit area (pressure) in the nip. Moisture removal is directly related to nip pressure. The pressure varies continuously in the machine direction within the nip, so it is sometimes convenient to use average nip pressure. This is calculated from the press loading per unit width divided by machine direction nip width, as in equation (2).

$$\text{Average Nip Pressure (kPA)} = \frac{\text{Nip Load (kN/m)}}{\text{m (of nip width)}}$$

or (2)

$$\text{Average Nip Pressure (psi)} = \frac{\text{Nip Load (pli)}}{\text{in. (of nip width)}}$$

A measure of nip width or uniformity of loading across the face of the press roll may be approximated by carbon paper impressions of the loaded nip with the press at rest. Pressure-sensitive films are also available for the same purpose and, combined with a density meter, can approximate pressure distributions in the nip. Total press loading can be estimated from pressure gauges connected to the pneumatic or hydraulic cylinders, or from electronic load cells.

Press Load Uniformity

Another press loading factor that appears to qualify as a primary variable in the wet press nip, uniformity of pressure applied to the sheet, has not yet been broadly understood or exploited. The uniformity referred to here is not of the type associated with variations in felt weight over a distance of a meter or of differences in pressure in the cross-machine (CM) profile due to an improper roll crown. Rather, the scale of uniformity being considered is over distances of a few millimeters or less.

The largest factor in nip pressure uniformity is the felt. Measurements have been made in the laboratory with pressure transducers rotating through a press nip. The transducer has a sensitive area on the order of 1 mm to 2 mm diam. Each revolution of the transducer through a felted nip produces a pressure profile with a different peak pressure while the residence time, or nip width, remains constant. The difference in magnitudes of these peak pressures, all at constant press loading, is astonishingly high. A peak pressure of 0.7 MPa (100 psi) is often followed by a peak pressure of over 3.4 MPa (500 psi) as illustrated in **Fig. 10.1.**

Beck *(8)*, in a rigorous study of nip pressure profiles, recorded a range of peak pressures from 1.5 MPa (226 psi) to 18.4 MPa (2677 psi) all at constant press load. Fekete and Waugh *(9)* have shown cross-sections of a 49 g/m^2 sheet pressed against a felt at 3.4 MPa (500 psi). When viewed at magnifications of 200 times to 300 times, the felt fibers appear to be very coarse relative to the sheet structure; large variations in pressure uniformity at this microscale seem plausible. Using aluminum-foil sandwiches of paper and blotters on a laboratory press, Sze *(10)* demonstrated the localized variability of felt intrusion into the sheet at the felt-paper interface. A photomicrograph of the foil is shown in **Fig. 10.2.**

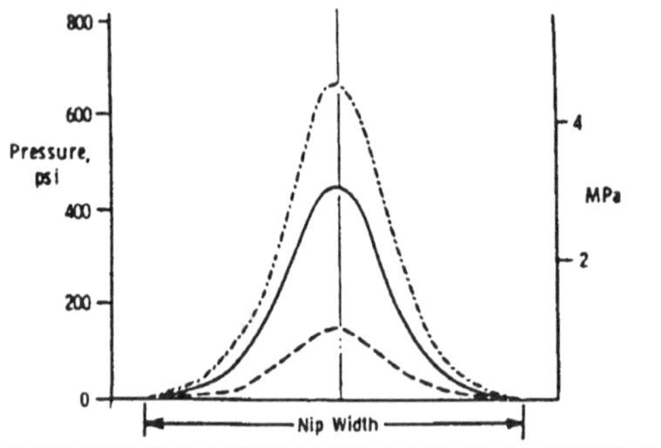

Fig. 10.1: Peak pressure variations

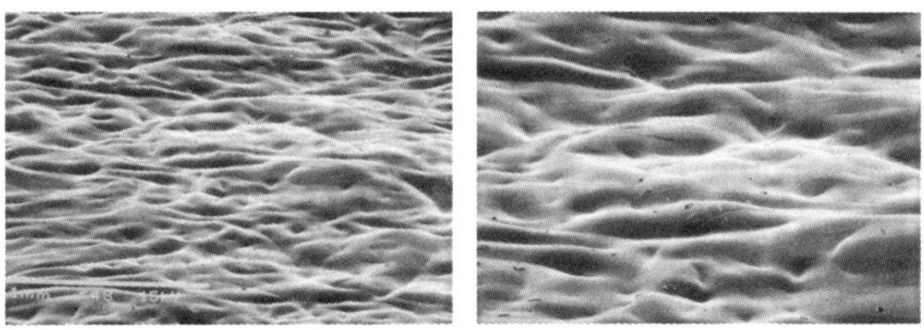

X48 X90

Fig. 10.2: Foil impressions of felt under loading

Of more interest here is whether the sheet water removal is affected by the felt nonuniformity or nonuniform pressure application on the microscale. Qualitative tests on laboratory presses at constant conditions suggest that fineness of the felt surface can control sheet dryness. If felt contact on the sheet is not uniform, then a localized high mechanical pressure at the felt-paper interface will result in localized high hydraulic pressures in the sheet. These localized hydraulic pressures cause the water between the fibers to migrate to lower-pressure areas in the sheet. This lateral transfer of water results in locally saturating the sheet in an adjacent area where it is difficult to effectively remove it to the press felt. Greater felt uniformity generally appears to increase outgoing press solids *(10)*.

This work is further supported by **Fig. 10.3**, where a very coarse felt produced a dryness of only 37.3% while a much finer structure produced 43.9% dryness. The felt patterns shown were obtained by carbon paper impressions of the felt at moderate pressure levels. Bliesner and MacGregor *(11)* prepared special baseless felts using a fiber diameter of 43 μm in one felt and 19 μm in another. The smaller-diameter fiber produced a dryness nearly 7% higher on a 50 g/m² sheet. Pressure uniformity in the press nip clearly is a primary variable.

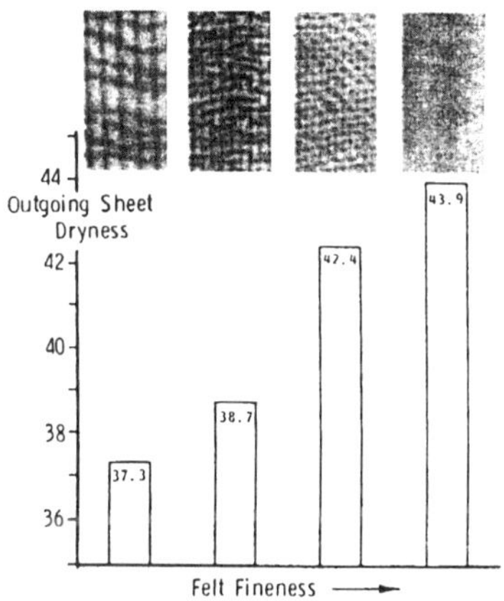

Fig. 10.3: Dryness vs. felt fineness

Machine Speed

Machine speed has a strong influence on water removal in the wet press. As speed is increased, water removal from the sheet will decrease with other conditions held constant. The more fundamental variable, however, is time for the water removal to occur. This generally is called nip residence time (NRT). Nip residence time is calculated by dividing the nip width by the machine speed (equation 3); it is normally expressed in milliseconds.

$$\text{NRT (msec)} = \frac{\text{Machine Direction Nip Width}}{\text{Machine Speed}}$$

$$= \frac{60 \times \text{Nip Width (mm)}}{\text{M/Min}} = \frac{5000 \times \text{Nip Width (in.)}}{\text{Ft/Min}} \qquad (3)$$

The constants are required to convert time and length units.

Machine speeds can be measured quite accurately, but the measurement of nip width is subject to greater variability.

Press Impulse

The effects of press loading and nip residence or dwell time historically have been treated as independent variables. Evidence suggests that the product of these variables unifies their influence on overall water removal in the press. Campbell *(12)* is the earliest investigator who equated dryness or increase in outgoing sheet consistency as being proportional to the product of pressure per unit area and duration of press application. Later, both Wahlstrom *(5)* and Schiel *(13)* used the pressure-time product in relation to wet press water removal. More recently, Beck *(8)* has succinctly summarized the definition, shown in equation (4).

$$\text{Impulse} = \int_{t_b}^{t_e} P(t)\,dt = P_{ave} \times (t_e - t_b) \qquad (4)$$

$$= \frac{\text{Press Load}}{\text{Machine Speed}}$$

Impulse is usually expressed as MPa-sec or psi-sec.

Where:

P_{ave} = average nip pressure, MPa or psi
$P(t)$ = pressure at any time t, MPa or psi
t_b = time profile begins (entering the nip), sec
t_e = time profile ends (leaving the nip), sec

This expression was used to obtain data on a laboratory press for a 60 g/m^2 bleached kraft sheet. The data shown in **Fig. 10.4** include a wide range of pressures and dwell times and suggest a single dryness-impulse curve.

The impulse applied to the sheet by typical two-roll press sections today is of the order of 0.013-0.02 MPa-sec (2-3 psi-sec). An increase in the press impulse by a factor of 10% would increase dryness out of the press by 8% or more; this makes press impulse an important primary variable.

The same data plotted against the log of the press impulse suggest that a straight-line relationship exists on the semilog plot through the limited range shown in **Fig. 10.5**. Obviously, the relationship holds true for a very wide range of residence times and pressures, or machine speeds and load combinations, in the range of 30% to 55% dryness. At some very high level, both pressure and time fail to have an influence.

Press impulse seems also to have value on even the very lightweight sheets, such as newsprint, where the effect of time would be expected to be less of a factor. Data collected on a repulped newsprint using a two-roll press are shown in **Fig. 10.6.** The dryness changes from near 30% at a press impulse of under 0.007 MPa-sec (1 psi-sec) to near 60% dryness at a press impulse of 0.07 MPa-sec (100 psi-sec), truly an amazing change in dryness.

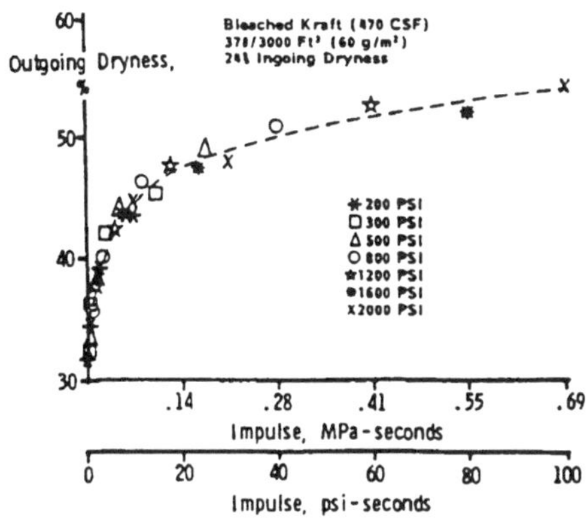

Fig. 10.4: Dryness vs. impulse, kraft

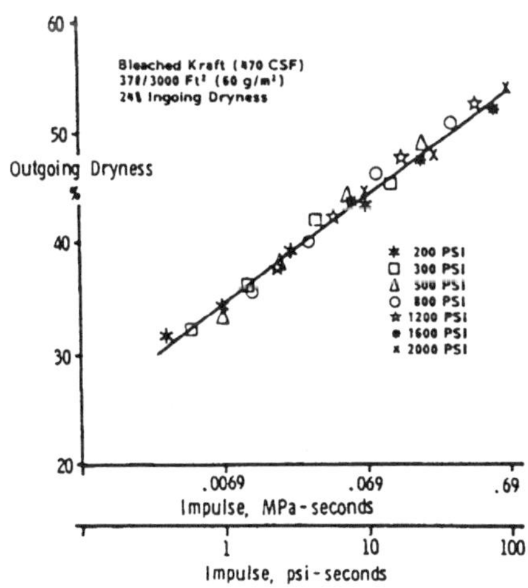

Fig. 10.5: Dryness vs. log impulse, kraft

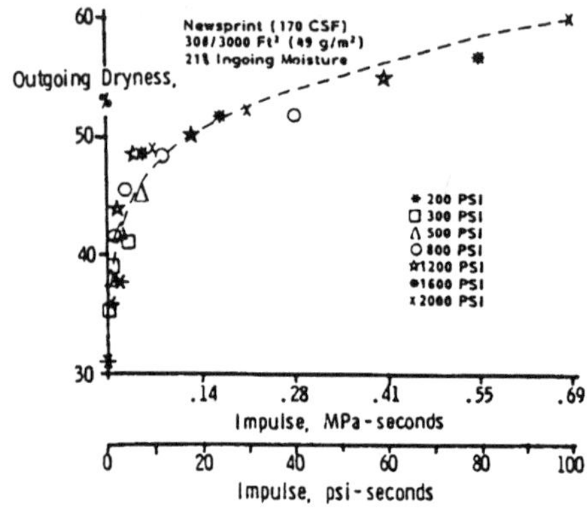

Fig. 10.6: Dryness vs. impulse, newsprint

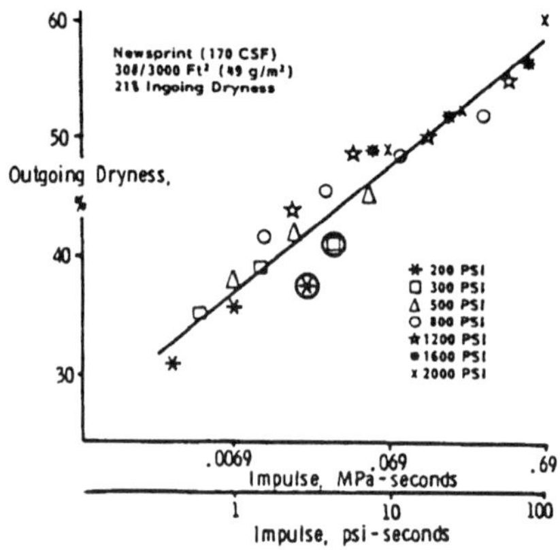

Fig. 10.7: Dryness vs. log impulse, newsprint

A semilog plot of the newsprint data is shown in **Fig. 10.7** to illustrate some differences from the somewhat heavier 60 g/m² kraft sheet of **Fig. 10.5**. The scatter of the data around the slope at the mid- to low-impulse range is believed to arise principally from 1% to 2% moisture rewet at the nip exit.

Many pilot studies have been conducted on linerboard furnishes using both roll and shoe-type wet pressing equipment. **Fig. 10.8** shows outgoing dryness vs. press impulse for four different linerboard furnishes pressed on a pilot shoe press. All pulps show similar trends but exhibit shifts in absolute dryness for the different furnishes. Roll presses typically have impulses of 0.012-0.02 MPa-sec (2-3 psi-sec) while current shoe presses attain values near 0.1-0.14 MPa-seconds (15-20 psi-sec).

When dealing with a product like linerboard, it does not seem to matter if the impulse is provided by a roll press or a shoe press. For example, an unbleached kraft stock at 27°C (81°F) that was difficult to dewater was pressed in both a roll-press nip and in a shoe; and the dryness values from the outgoing nip show a continuous curve (**Fig. 10.9**). Due to the additional water removal using long-nip presses, linerboard and corrugating medium grades appear to be the most popular application of this technology.

Linerboard sheets, where the compaction is controlled primarily by the outflow of water, are said to be flow controlled. In some of the lightweight sheets, compaction is controlled more by the compression of fibers and less by water flow; these are said to be pressure controlled.

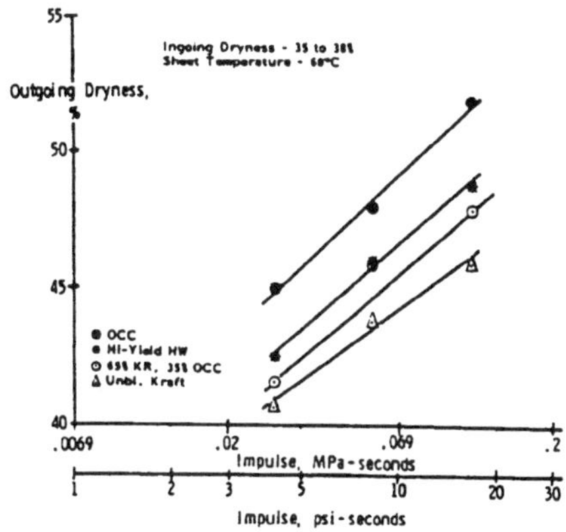

Fig. 10.8: Dryness vs. impulse, linerboard

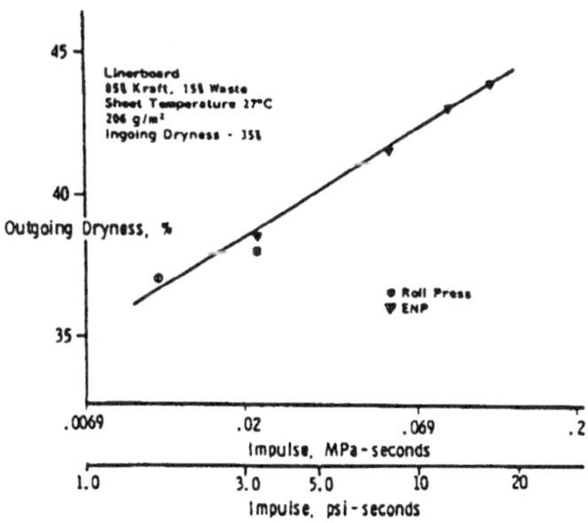

Fig. 10.9: Dryness vs. log impulse, roll and extended nip press

Sheet Temperature

The temperature of the sheet as it is being pressed can have significant effects on water removal and thus qualifies as a primary variable *(14, 15)*. As the temperature in the sheet is raised both surface tension and viscosity of the water decrease, which lowers the resistance of water movement through the sheet and to ultimate removal into a felt at the press. Typical data are shown in **Fig. 10.10.** At 33% in-going moisture level the dryness has increased from 40% to near 45% with an increase of temperature from 27°C (81°F) to 82°C (180°F), or 1% dry for each 11°C (52°F) increase. It is important to emphasize that the effects of temperature are independent of the other press variables and that individual benefit can be achieved from each. See (16) and **Fig. 10.11.**

In general, heating the sheet can save energy (17). Steam showers on the press and heated press rolls are both used to heat the stock.

Some propose that the increase in temperature softens the fiber and allows for reduced compressive resistance in the nip leading to higher exit solids from the nip. While heating the sheet is effective in increasing press dewatering, it is not proven that the fibers actually soften in the temperature range used in conventional pressing.

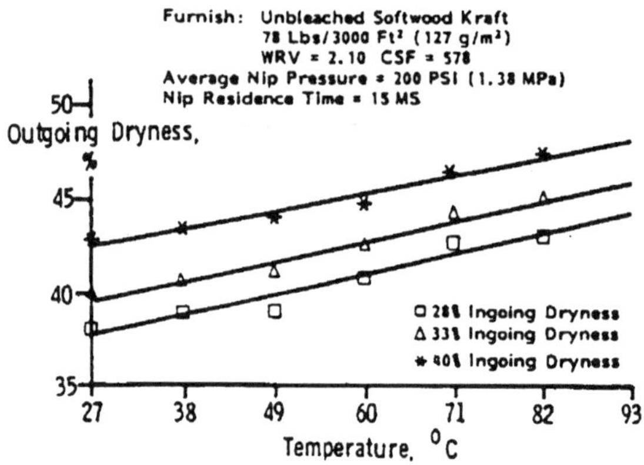

Fig. 10.10: Dryness vs. sheet temperature

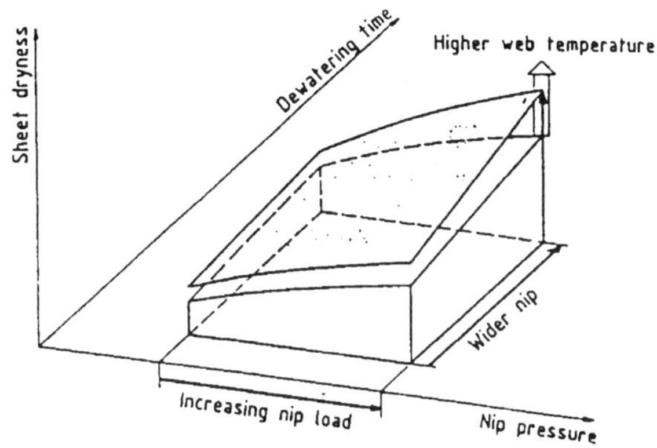

Fig. 10.11: Interrelationship of process parameters. (From Ref. 16)

Polymers classically soften above their glass transition point, T_g. While even the existence of a thermal softening temperature for cellulose is widely debated, if it does exist it is in a range (200-260°C, 392-500°F) far higher than that achieved on a conventional heated press section. Similarly, the T_g for dry hemicellulose and kraft lignin is between 150°C to 250°C (302°F to 482°F) and 125°C to 200°C (257-392°F), respectively *(18)*. However, moisture plays an important role of plasticizer, which effectively reduces the T_g of both the hemicellulose and lignin polymers to a lower temperature range *(19)* although still above the range commonly encountered in a heated paper machine press. From the experiences gained in laboratory press drying and impulse drying experiments, it is generally agreed that the sheet webs at 30% to 60% moisture generally dynamically soften (i.e., produce instantaneous flow) at temperatures between 100°C to 150°C (212°F to 302°F). Evaporative cooling of a heated sheet at atmospheric pressure would always keep its effective temperature from ever nearing 100°C (212°F).

In-going Sheet Dryness

In-going sheet dryness is known to have a large effect on the outgoing sheet dryness. Smart *(20)* reported results of pilot machine testing using a 66 g/m^2 kraft sheet at 610 m/min (2000 ft/min) at a second press condition. At the 35% to 40% dryness level, a near linear relation of approximately 1% change in outgoing dryness was found for each 2% change in in-going dryness.

Busker *(21)* reported the effects of in-going dryness over a broad range, 24% to 38%, and found the outgoing sheet dryness was changed by as much as 5% through that range (**Fig. 10.12**). These values are typical of the range of values usually found, 3:1 to 2:1 change in the ratio of in-going to outgoing dryness. In-going sheet dryness is a major factor in determining outgoing dryness and is a primary variable; it is more often viewed as one of the determinants of water load to the nip. It is established by prior processing rather than being a controlled variable.

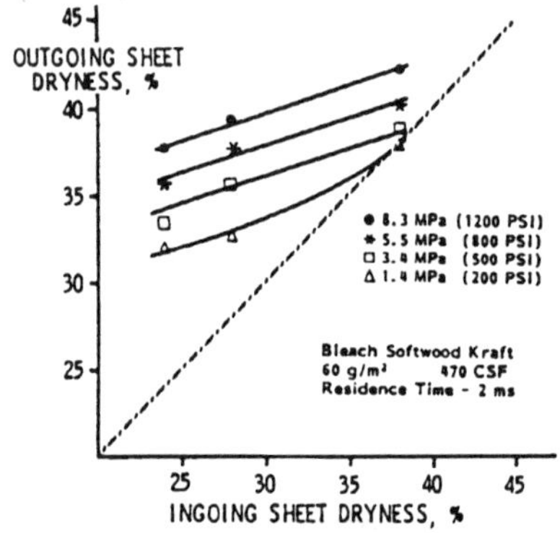

Fig. 10.12: Effect of in-going sheet dryness

Basis Weight or Grammage

Sheet basis weight or grammage is another factor that alters the water removal load at the nip and is dependent on the product line being manufactured. Ain et al. *(22)* observed a change in outgoing sheet dryness from about 43% to 39% when changing only basis weights from 45 g/m^2 to 120 g/m^2. Busker *(21)* also measured dryness at constant conditions while varying grammage from 60 g/m^2 to 163 g/m^2. This resulted in a maximum change of outgoing sheet dryness from 44% to 37% dryness at 3 MPa (435 psi) and is shown in Fig. 10.13. Andersson and Back *(14)*, using unbleached kraft in a press simulator, found a change in dryness of 45% to 41% for sheets of 60 g/m^2 to 240 g/m^2 basis weight at 70°C (158°F). Variability in sheet grammage is a major factor in water removal, albeit a predetermined and given one.

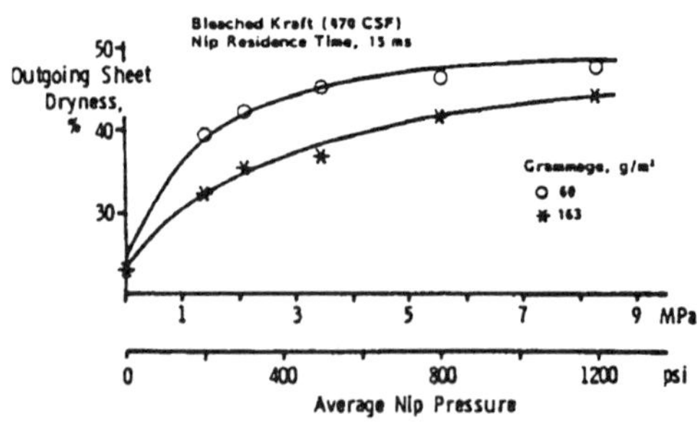

Fig. 10.13: Effect of basis weight on grammage

Furnish Properties

Although we stated that we would attempt to establish the steady-state factors in a wet press that influence water removal, the furnish properties are never at a steady state although they are primary variables in their impact. Furnish variability begins with diversity of tree species, age, and condition of the fiber in the chip entering the pulping process. Variability in chip thickness before cooking and the natural instability of the pulping and bleaching operations impart variability in the pulp not present in the native fiber. Freeness, which is largely controlled by a few percent of fines material, is inadequate to clearly predict the sheet consolidation and press dewatering performance.

As the sheet is consolidated on the machine wet end, comparatively minor forces are exerted on the fibers to compact the sheet. By contrast, the wet press represents a significant mechanical compaction. The resistance of individual fibers to collapse are forces which oppose sheet consolidation.

A wet mat's resistance to being compressed is related to the individual and distributed characteristics of the fibers in the sheet. When the contact junction of two fibers is transversely compressed in the nip, stresses are localized in the region of the fiber-fiber crossing. The load required to initiate fiber buckling under the instantaneous compression in a press will depend on the wall thickness and the outside fiber diameter. Luce *(23)* developed the fiber shape factor shown in **Fig. 10.14**, which estimates the relative transverse buckling load required to promote an individual fiber's collapse. Consequently, it is not surprising that even within the same species changes in the proportion of earlywood and latewood fibers, thin-walled juvenile wood fibers, or the chemical yield of the fibers will influence a mat's response under dynamic compression. The extent to which the mat is compressed will directly control the amount of water expressed from the sheet.

However, if hydraulic drag of water movement inside the sheet increases, as might occur with an increase in fiber fibrillation and fines, then water removal will be retarded. Slight changes in the ease with which water is able to be removed out of a sheet will depend on whether the sheet deformation is pressure controlled or flow controlled.

Carlson *(24)* and Caulfield and coworkers *(25)* liken the compression characteristics of a wet sheet in a press to Kelvin elements shown in **Fig. 10.15**. When a sheet is pressure controlled, little resistance is offered to the flow of water out of the mat and the mat elastically rebounds after the sheet emerges from the nip. This rapid rebound of such furnishes probably influences their rewet gain. However, if the fiber deformation rate is viscously dampened and resistance to fluid flow within the sheet is increased due to high surface area on the fibers, higher basis weight, or more swollen components, then pressing this sheet will be more flow controlled. Flow-controlled pulps are thought to behave like the viscous element in **Fig. 10.15** where the sheet resists both deformation upon entering the nip and elastic expansion once out of the nip.

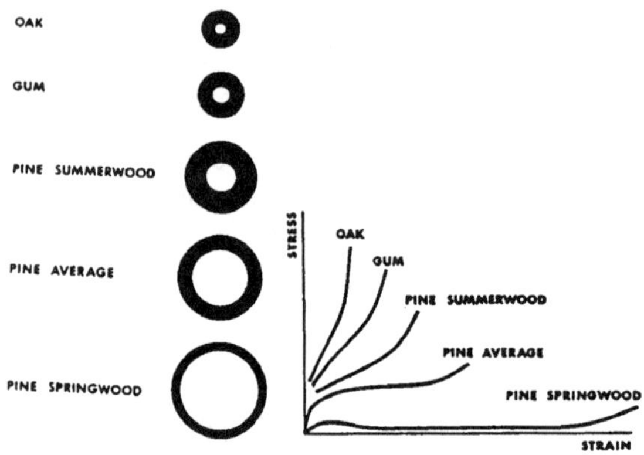

Fig. 10.14: A comparison of cross-sections and stress-strain behavior of fiber models (From Ref. 23)

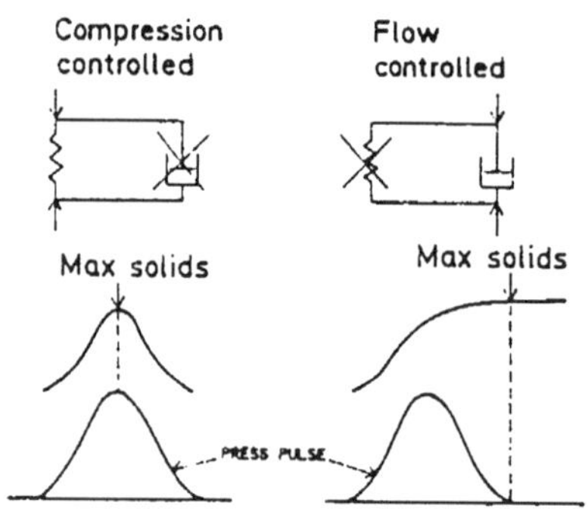

Fig. 10.15: Deformation of Kelvin elements when either the spring or the dapshot determines the compression response

Simply knowing the freeness of different furnishes does not adequately estimate how the stock will either dewater or consolidate in wet pressing. **Fig. 10.16** shows the scatter in sheet press dryness vs. the Canadian Standard Freeness (CSF) for about 60 furnishes having a broad spectrum of pulp types and refining conditions.

Busker *(21)*found a change of about 15% dryness at typical wet press conditions on a bleached kraft sheet when refined from 720 mL to 200 mL CSF (**Fig. 10.17**). These data show the large effect of refining on dryness at commercial speeds in the time range of 2 - 10 msec NRT, and also illustrate the flow of rewet water and the time dependency of rewet water at residence times above 50 msec.

A better correlation has been found between the pulp's water holding capacity (determined as its water retention value [WRV]) *(26)* and outgoing dryness from the wet press. However, the results can be influenced by changes in fines content and other factors. **Figure 10.18** shows the correlation of dryness to WRV for the same furnishes as in **Fig. 10.16.**

Research efforts continue to seek simple, reliable methods of measuring one or more characteristics that control flow inside and outside the fibers. Perhaps the ultimate method is a direct measurement of water removal using a bench-type press to determine a "pressability index" for each furnish.

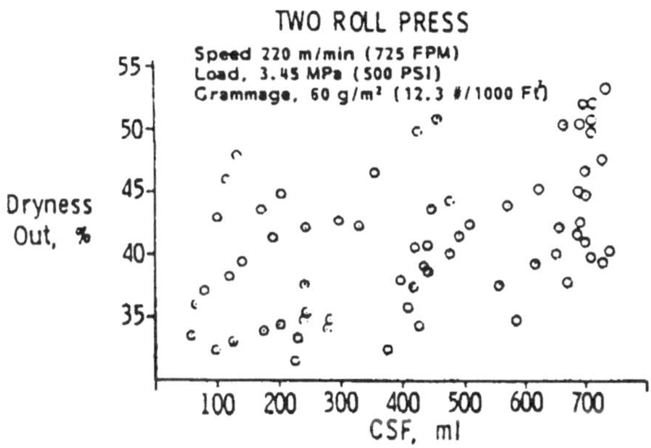

Fig. 10.16: Dryness vs. freeness for various furnishes

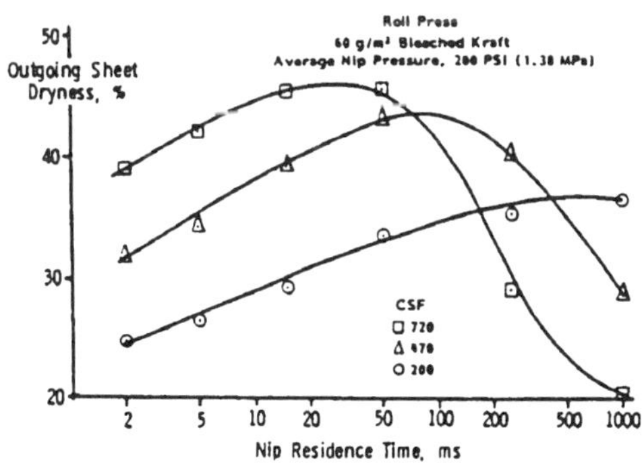

Fig. 10.17: Refining effects, kraft

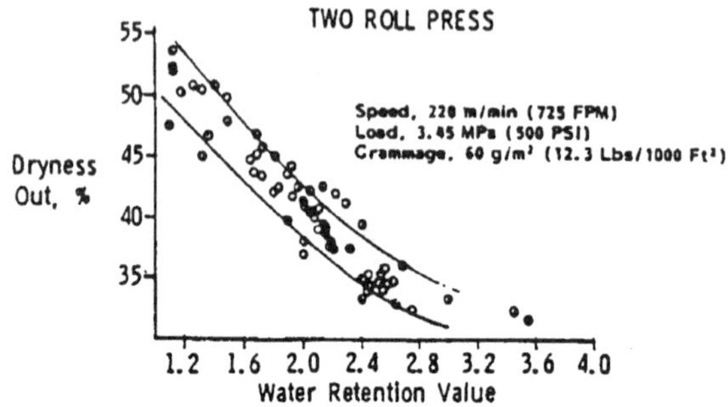

Fig. 10.18: Dryness vs. retention value

Felting

The wet felt is the receptor that acts as an absorptive interface between the press roll and the sheet. The considerations on felt surface uniformity mentioned earlier can have a dramatic effect on sheet water removal.

Felt compressive properties are controlled by felt condition and design. The elasticity of the felt exiting the nip will control the rate and even the direction of water transfer at the sheet-felt interface. The difference in void volume between the compressed and relaxed conditions is a first approximation of the felt's absorptive capacity. Perhaps more important is the felt's void volume at the most compressed state.

Double felting has been an effective method of increasing water removal in the press. Perrault *(27)* reported the results of a TAPPI Pressing and Drying Committee survey on experiences with double felting on linerboard machines. Many of the results indicated an increased dryness of 4% or more. Double felting is an important, and thus a primary, variable on heavyweight sheets. It is less effective on lightweight sheets and is less valuable for such applications.

Rewet

Rewetting (rewet) is probably the most controversial aspect of wet pressing. It is an effect rather than a variable in the strict sense and probably should be treated as negative water removal rather than as a wet press variable, but most authors include it as a variable.

Wahlstrom *(5, 6)* has suggested that the quantity of water going back into the sheet (rewet) in the expanding zone of the nip varies from 3 - 35 g/m^2. Bliesner and MacGregor *(11)* and Thorne *(29)* have measured rewet external to the nip by wrapping the felt against the sheet after the pressure zone. They measured values of 7-37 g/m^2. Wheeldon and Hoyland *(28)* used chemical tracers to measure rewet at low speeds and values up to 45 g/m^2 were measured. This return of water to the sheet can have a significant effect on the solids content.

Jaavidaan *(30)*, in bench tests on bleached softwood kraft, found that rewet is independent of press impulse for a constant amount of water removed during compression. Approximately 30% of the water removed during compression was returned to the sheet as rewet.

Post-nip rewet is a second important aspect of rewetting. If sheet and felt are left in contact after leaving the nip zone, a post-nip rewet can occur. Even though rates of transfer of water back to the sheet may be falling off, the total water transferred can be large if a long rewet time is allowed. For example, a contact length of 1 m following the nip on a newsprint sheet could result in a post-nip rewet of 5% based on the data presented in **Fig. 10.19** and shown in **Table 10.2.**

Table 10.2 Consequences of a rewet

Grammage g/m²	Speed m/min	Rewet time ms	Rewet g/m²	water MR	Change in sheet dryness at 40% level	D
Rewet in 20 mm expanding nip (based on 650 CSF data).						
15	1500	0.8	4.0	.27	40 - 36.1	3.9
49	900	1.3	5.0	.10	40 - 38.5	1.5
130	600	2.0	6.3	.05	40 - 39.2	0.8
205	450	2.7	7.5	.036	40 - 39.4	0.6
Rewet in 1000 mm post nip (based on 470 CSF data)						
49	900	66.7	17.5	.36	40 - 35.0	5.0
205	450	133	28.3	.14	40 - 37.9	2.1

Even a heavyweight sheet like linerboard can lose 2% or more moisture given the contact time after the nip shown in **Table 10.2.** Baughman and Mattila *(31)* documented a change in post-nip contact time on a paper machine resulting in a 2.9% improvement. Post-nip rewet can be a primary negative effect, but it can almost always be avoided by careful design.

C. SECONDARY VARIABLES

A secondary variable is defined here as one that shows a promise of improving press dryness by less than 4%, even though large improvements over state-of-the-art levels of that limit could be achieved. These variables are important because their combined improvements can have significant benefits on water removal. Because they are secondary, they are less apt to be productive areas for influencing large gains in water removal.

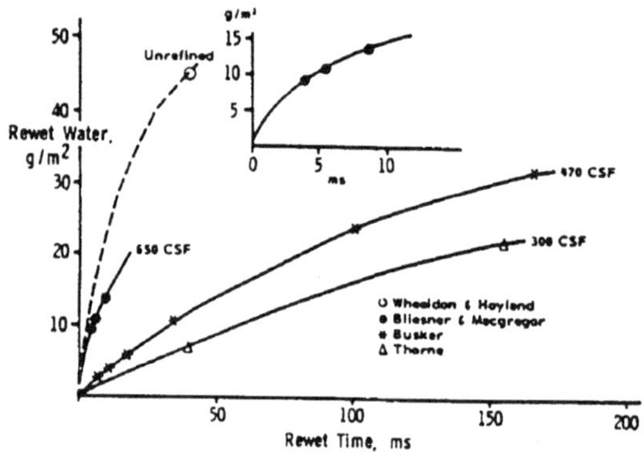

Fig. 10.19: Rewet vs. time

Felt Flow Resistance

The role of new felt flow resistance was reviewed by Bliesner *(32),* who indicated that at a compacting pressure of 2.4 MPa (350 psi), the hydraulic pressure drop across an unrefined softwood kraft sheet is 30,000 times the pressure drop across a modern wet felt at the same condition of transverse flow. This was also confirmed by Kershaw *(33),* who carefully documented felt flow resistance and pressure drop across new felts under a compression of 2.4 MPa (350 psi). The recorded pressure drops in the z-direction on new felts were less than 1 kPa (0.14 psi) at flow rates equivalent to water flow from a lightweight sheet.

Chang *(34)* made direct measurements of hydraulic pressure at both the felt surface and at various weight paperboards. The maximum hydraulic pressure measured at the face of the felt did not exceed 0.2 MPa (29 psi) (and was thought to be actually much lower under dynamic conditions) while a heavy board sheet reached levels of hydraulic pressure over 3 MPa (435 psi). In this particular experiment the pressure drop across the felt was less than 7% of that across a heavy board sheet. In conclusion, initial felt flow resistance is very low and no significant improvement in sheet dryness is to be achieved by lowering the initial flow resistance of wet felts.

However, felt flow resistance increases significantly during use on the machines to the point where the felt must be removed. The research emphasis should be on the change of felt flow resistance during aging.

In-going Felt Dryness

Many articles on wet pressing have started with the assumption that the in-going felt moisture ratio will strongly influence the exiting sheet moisture ratio in the wet press. This assumption has a strong appeal. The experimental evidence suggests in-going felt moisture is very limited in controlling outgoing sheet moisture. Several investigators *(35, 36, 37, 38)* have studied this variable at commercial machine speeds and all agree the effect is less than 1% of dryness (**Fig. 10.20**). Only two exceptions to this position have been observed:

1. Very-low-speed laboratory experiments *(39, 28)* below 30 m/min (100 ft/min) where excessive rewet was able to occur and the available water in the felts and vents could influence the magnitude of the water going back into the sheet in the expansion zones.

2. Cases of double felting where samples remained in contact with the felt after the nip.

Perhaps it is not surprising that the felt in-going moisture ratio has only a small effect on exiting sheet moisture if the pressures required to reach saturation of the felt in the nip are reviewed. Luciano *(40)* found that for an in-going felt moisture ratio of 0.6 (not uncommon in machine operation) that a batt-on-base felt saturates with an applied pressure of less than 0.2 MPa (30 psi). A batt-on-mesh felt reaches saturation in the nip at 1.24 MPa (180 psi). Many press nips operate at average nip pressures greater than 1.4 MPa (200 psi), so Luciano's finding suggests that most felts will reach saturation in the press nip. When conventional conditioning equipment is used, the entering felt moisture will seldom be reduced to a level that avoids felt saturation at the nip.

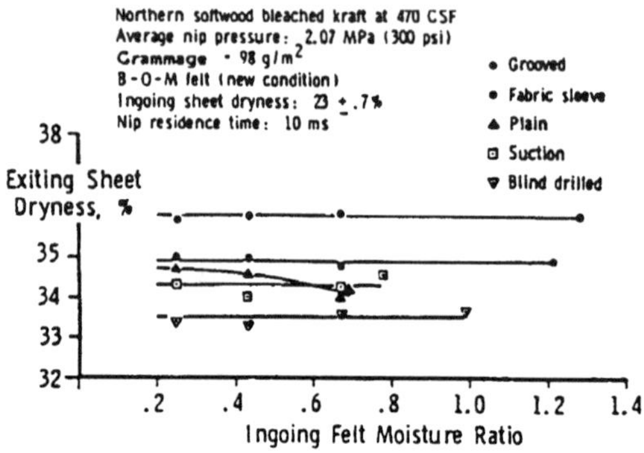

Fig. 10.20: Sheet dryness vs. felt moisture

When a felt saturates in the press nip, it is necessary to design into the press the capability to remove water before the nip, or by venting at the nip. This will facilitate the maximum transfer of water from the sheet to the felt and minimize sheet rewetting. Otherwise, felt saturation can have a dramatic effect on outgoing sheet dryness.

Limited investigations on plain press rolls *(32, 37)* suggest that roll venting is effective in removing free water when the felt reaches saturation. Consequently, with an effectively vented press roll, high incoming felt moisture contents exhibit less than 1% change in sheet dryness. The venting of nips and conditioning of felts (which includes felt cleaning, decompacting, and maintaining the ease and uniformity of water removal) will help to minimize the negative water removal effects of felt performance that can plague the papermaker.

Pressure Profile Shape and Roll Hardness

The pressure profiles of the roll press and the shoe press are fundamentally different, as shown in Fig. 10.21. However, when the area under the profile is the same, the press impulse is considered equivalent. If both press speed and loading are constant, it does not matter what shape the pressure profile takes, the press impulse (the integral of pressure over the time interval) is constant and the dryness from the press will be the same for any given furnish (see Fig. 10.9).

Andersson and Back *(41)* used a dynamic press simulator to press a 120 g/m² unbleached kraft sheet with three different nip profile shapes (3, 6, 12 msec; 8.2, 3.8, 1.4 MPa peak pressure). The impulse was approximately constant for the first two conditions and the outgoing sheet dryness was nearly identical for the two pulse shapes. The third condition (12 msec) is a lower impulse value and a lower dryness value was recorded.

Changes in roll cover hardness or felt compressibility would not likely have significant effect on a flow-controlled furnish such as in linerboard. These changes only contribute to a minor reshaping of the pressure profile and, if press loading and speed are not changed, the impulse will essentially remain the same. While this effect should not be generalized across the broad spectrum of paper and board grades, certain furnishes such as pressure-controlled stocks may exhibit a small benefit with a change in roll hardness. While each case must be individually examined, available data suggest that roll cover and felt hardness will affect the profile shape but should be considered secondary variables.

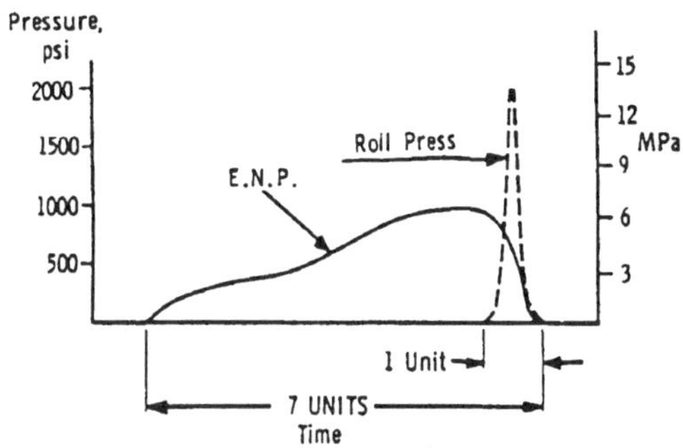

Fig. 10.21: Nip pressure profiles

Nip Venting

The original paper machine press, or plain roll press, was operated at low machine speeds. Excess water squeezed from the sheet and felt had sufficient time to run down the in-going face of the bottom press roll and escape from the nip. As machine speeds increased, it became important to provide a short flow path for the water to escape from the felt. Suction rolls were used first to provide a storage area for the water removed by compression in the nip. Since then, several approaches have been used to provide for venting or momentary storage of the removed water: additional water-holding fabrics were run underneath the felt, or grooves were cut or holes were drilled into the face of the felted press roll.

The grooved roll has been shown to be very effective in providing venting and seems to offer slightly better water removal than other types of venting (see **Fig. 10.20**). The plain press can be very good when the felt water load is very low, i.e., less than 0.4 in-going felt moisture ratio. As speed goes up and water load increases, the plain roll felt quickly becomes saturated and water removal performance degrades rapidly. The likelihood of sheet crushing increases, also.

Nip pressure uniformity can be influenced by the venting in the press roll. Busker and Gordon *(38)* found that venting with a large groove span of unsupported felt, 3 mm (.118 in.), produced a lower dryness value than venting with a shorter span, 0.5 mm (.02 in.). A difference of up to 2% dryness was measured. Use of shorter spans in the venting, and felts designs that provide better bridging of the vent opening, minimize the effect of vent pressure nonuniformity.

D. OTHER FACTORS

Water removal through progressive roll press nips becomes smaller at each nip. A first press position commonly increases consistency from an in-going value of 15% to 20% to an outgoing value of 28% to 35%. Of course, first press water removal is dependent on many variables: grammage, impulse, and others. First nips are often wide, operated at low press loading, and sometimes employ open, bulky felts. Double felting is frequently used for enhanced water removal. The limit of press loading is often set by water removal factors which contribute to crushing the sheet in the first press nip. There is some speculation that water removal can be enhanced in the first press by actually operating under very mild sheet crushing conditions, but this can have negative effects on the sheet's strength properties.

The press in the second position will often increase the consistency to the range of 32% to 38%. The nip loading will be higher and a harder roll will be used. The third press position will similarly be of higher press loading, which may result in increased consistency in the 40% to 45% range. It is important to have an increasing nip pressure in successive press nips. Trials on pilot machinery have shown that passing the sheet through a press nip with lower nip pressure than prior nips will resaturate the sheet with water from the felt and the outgoing consistency will be decreased.

The shoe-type press can be used in any press position. The most frequent application has been following a roll-type first press. There are now several applications where the shoe press is used in both a first and a second press position. The consistencies out of the shoe press are much higher, reflecting the greater impulse. When a shoe press is used, it is not uncommon on some of the easily dewatered furnishes to achieve first press consistencies well above 50%.

Press position alone is not known to have any significant role in water removal.

E. SUMMARY

Any attempt to classify the importance of the primary and secondary factors that control water removal in a wet press can be challenged easily by either altering the conditions or by comparing these conclusions with an isolated set of data. **Table 10.1** summarizes and categorizes the current view of wet pressing variables as presented in this chapter. Meaningful clarification will arise from the cooperative efforts of engineers, researchers, and designers when their findings lead to the development of advanced diagnostic techniques, improved control and detection devices, and the application of new press and felt materials or designs. Studies which assist us in more completely appreciating the dominance and roles of some of the currently recognized variables, and the subtle influences of those variables not yet fully appreciated or identified, will lead our industry to a better understanding of the factors controlling water removal processes in the paper machine wet press.

It is our sense that there are three primary variables where application and innovation would be very productive. Increasing sheet temperature, time and pressure (press impulse) can cause large and more or less independent gains in water removal from the sheet. The support by both industry and government of fundamentally oriented science and expanded engineering studies on water removal in the wet press is needed to achieve the important goal of higher production with lower energy use while improving product quality.

Portions of this chapter originally published in Pulp & Paper Canada 85(6): T138 – 147 (1984). Reprinted with permission from the Technical Section, CPPA.

References

1. Robertson, G. J., The Evaluation and Improvement of Press Section Performance: Some Practical Aspects, APPITA 30(1): 62(1976).

2. Bliesner, W. C., Sheet Water Removal in a Press: Time to Review the Fundamentals, Pulp Paper 52(10): 75 (1978) and 52(11): 76(1978).

3. Cox, C. J. and Robertson, G. J., The KMW Press Simulator, a Tool for Pressing Studies, 1981 APPITA General Conference, 1981.

4. DeCrosta, E. F., Toward a Unified Theory of Pressing, CPPA Annual Meeting Technical Section Preprints, 1981, p. B105.

5. Wahlstrom, P. B., Our Present Understanding of the Fundamentals of Pressing, Pulp Paper Can. 70(19): 76(1969).

6. Wahlstrom, P. B., New Developments and New Insights in Water Removal by Pressing, EUCEPA Paper No. 28, London, 1979.

7. Bullerwell, A. C., Sutherland, P. H., and Krotchko, D. J., Development of an Engineered Nip, TAPPI Papermakers Conference Preprints, TAPPI PRESS, 1980, p. 105.

8. Beck, D. A., The Dynamic Properties of Paper Machine Wet Felts in a Press Nip, TAPPI 1980 Engineering Conference, Book I, TAPPI PRESS, 1980, p. 181.

9. Fekete, E. Z., and Waugh, R. W., Flat Fiber Felts Reduce Break-In Time, Improve Press Performance, Pulp Paper 55(2): 124(1981).

10. Sze, D. H., Measuring Wet Press Felt Pressure Uniformity and the Effects on Sheet Solids, TAPPI Engineering Conference Proceedings, TAPPI PRESS, 1985, p. 149.

11. Bliesner, W. C. and MacGregor, M. A., How Important Is Rewetting in Wet Pressing, Tappi J. 59(5): 114(1976).

12. Campbell, W. B., The Physics of Water Removal, Pulp Paper Can., Convention Issue, 103(1947).

13. Schiel, C., Optimizing the Nip Geometry of Transversal Flow Presses, CPPA 2nd International Water Removal Symposium, Mount Gabriel, P.Q., 1968.

14. Andersson, L. and Back, E., The Effect of Temperature up to 90°C on Dewatering of Wet Paper Webs, Evaluated in a Press Simulator, TAPPI Engineering Conference Proceedings, TAPPI PRESS, 1981, p. 311.

15. Royo, M. and Thorp, B., The Effect of Direct Steam Heating on Water Removal, PIMA 63(9): 38(1981).

16. Grant, R., Modern Press Sections: Tracing Trends for the Future, Pulp Paper Int., Vol 30 (52)(1988), p. 55.

17. Paulaporo, H., Pressing and Drying, A Revolution Ahead? Proceeding XXIII EUCEPA Conference, Harrogate, U. K., vol. 1, 1988, p. 246.

18. Back, E. L., and Salmen, N. L., Glass Transitions of Wood Components Hold Implications for Molding and Pulping Processes, Tappi J. 65(7): 107(1982).

19. Horn, R. A., Bonding in Press-Dried Sheets From High Yield Pulps, Tappi J. 62(7): 77(1979).

20. Smart, F. R., Water Removal Performance on a Grooved Second Press, Part I, Paper Tech. Ind., 16(3): 91(1980).

21. Busker, L. H., Wet Press Water Removal Over a Wide Parameter Range, Paper Tech. Ind. 16(3): 172(1975).

22. Ain, R., Fleischer, T. B., and Smart, F. R., A Study of the Effect of Drainage Load on Felt Water Removal Performance, APPITA 29(3): 180(1975).

23. Luce, J. E., No. 21: Transverse Collapse of Wood Pulp Fibers, Fiber Models, TAPPI Special Technical Association Publication; No. 8: The Physics and Chemistry of Wood Pulp Fibers, 1969, p. 278.

24. Carlsson, G., Linstrom, T., Norman, B., Some Basic Aspects of Wet Pressing of Paper, Journal Pulp Paper Sci. 9 (4): 101(1983).

25. Caulfield, D. F., Young, T. L., and Wegner, T. H., The Role of Web Properties in Water Removal by Wet Pressing, Tappi J. 65(2): 65(1982).

26. TAPPI Useful Method No. 256.

27. Perrault, R. D., Experience with High pli Double-Felted Presses on Linerboard Machines, Tappi J. 64(9): 80(1981).

28. Hoyland, R. W. and Wheeldon, J. B., Measurement of Fluid Transfer Within a Press Nip, Paper Tech. Ind. 21(5): 170(1980).29. Thorne, J. T., The Effects of Capillary Rewet on Wet Pressing Operations, Master's Thesis in Chemical Engineering, Univ. Maine at Orono, May 1981.

30. Jaavidaan, Y., Ceckler, W., and Thompson, E., Rewetting in the Expansion Side of Press Nips, Tappi J. 71(3): 151(1988).

31. Baughman, S. E. and Matilla, L. J., The Results of Opening a Second to Third Press Draw. TAPPI Engineering Conference Proceedings, TAPPI PRESS, 1978 p. 81.

32. Bliesner, W. C., Sheet Water Removal in a Press: The Role of Wet Felt Properties, Pulp Paper 52(11): 76(1978).

33. Kershaw, T. N., The Three Dimensions of Water Flow in Press Felts, Tappi J. 55(6): 880(1972).

34. Chang, N. L., Dynamic Compression of Handsheets, TAPPI Engineering Conference Proceedings, Book I, TAPPI PRESS, 1978, p. 96.

35. MacGregor, M. A., Duplication of Pilot Pressing Studies Using a Laboratory Web Former, Tappi J. 58(6): 116(1975).

36. Eiderbrant, A. and Swanberg, O., New Fabric Press Development Greatly Simplifies Operation, Paper Trade J. 149(26): 32(1965).

37. Osterberg, L., On the Felt Dryness Content, Wochenblatt Fur Papierfabrikation 97(21): 897(1969).

38. Busker, L. H. and Gordon, A. L., A Performance Comparison of Plain and Various Transverse Flow Presses, Tappi J. 64(2): 103(1981).

39. Wiseman, N., The Effect of Felt Properties on the Pressing of Paper, Pulp Paper Can. 77(9): 29(1976).

40. Luciano, W. A., A Practical Model for Felt Compressibility. Paper presented at the 68th Annual Meeting, CPPA, Technical Section, Montreal, 1982.

41. Andersson, L. I. and Back, E. L., Improvements in Dewatering at Increased Pressing Temperatures, a Press Simulator Evaluation, Tappi J. 65(7): 75(1982).

42. Busker, L. H. and Cronin, D. C., The Relative Importance of Wet Press Variables in Water Removal, Pulp Paper Can. 85(6): T138(1984).

Chapter 11

Practical Press Operation

A. INTRODUCTION

The maintenance of efficient press performance is contingent on many factors. Press performance depends on stock and sheet conditions entering the press and also relates to sheet quality requirements long after the sheet leaves the press.

Soft stock, with high water retention value (WRV) or low freeness caused by some fault in pulp manufacture or stock preparation, will result in poor press operation. A wetter sheet leaving the press causes slack draws which cannot be tightened without breaking the sheet. The sheet can wrinkle and fold over at the edges; the wet sheet will crush very easily in the press nips; and the press felt will rapidly fill and thus lose its water carrying capacity.

Common problems with the incoming sheet include:

- Poor sheet formation
- Streaks of poor formation on the wire
- Low sheet consistency
- Soft stock condition
- Incoming draw slack and uneven
- Contaminants in the stock

A press on a particular machine will develop a certain pattern of performance. This pattern is a benchmark to determine normal press performance and felt life. The press pattern will change with changes in press felt design, press conditions such as nip loadings, press roll plastometers, vacuum box positions, or vacuum levels and roll wear.

B. OPERATING OBSERVATIONS AND PROCEDURES

Press Log

During his shift the machine tender will make routine checks of the press noting the press performance. He will enter his notes and observations in an operating log for that machine. Develop the format of the log with considerable care. All information necessary to evaluate the performance of the press should be available. The machine tender will note, in addition to these observations, any adjustments such as draw, loading, shower position of pressure, etc., which he made. He should list reasons for the changes, also, and should determine and note the cause of any press breaks. It is often possible to locate the torn end of the sheet and to determine the cause and location of the break by examining the end. The sheet will normally tear back from the point of the break. This gives a clue at the reel how and where the break occurred.

Listed below are items which the log should contain:

- Uhle box vacuums
- Press loading front and back in PLI
- Press drive load or torque
- Roll suction vacuum
- Shower pressures on press felt cleaning and roll cleaning showers
- Press felt life
- Press felt tension and trade line shape
- Draw
- Sheet temperature
- Press exit moisture, if known, with 2 $\sum$ moisture variation
- Doctor blade pressure
- Vacuum pump seal water flow
- Vacuum pump operation including load
- Vacuum box position in suction rolls

Statistical Process Control (SPC)

Many mills are implementing new methods of logging information. They use statistical process control (SPC) to monitor the normal variability of the press performance. Press exit moisture can indicate when changes to the system are required; this means the process is getting out of control. This analysis is also useful when analyzing new press felts, rolls, and other equipment. Computers also can be used to record operating logs. Management and operators can then quickly analyze data over several days or months of operation. Numbers out of control limits are reported and acted on.

The use of an SPC chart of X-bar (average) and R (range) define the limits of normal control for the press (3 sigma limits). Any readings outside these limits tell the operator that something has definitely changed and he must search to find it before it shows up in the finished sheet. Many rules exist for analyzing the charts, and their use can prevent off-quality production. SPC prevents the changing of machine controls when a change is unnecessary, and conversely, not making a change when it should be changed. This prevents erratic operation due to excessive changes to the process.

Key parameters to chart include headbox freeness, couch vacuum, press moisture, or even draw. The computer logging systems can take input from the operation and output the SPC charts on the video display screen.

Press Performance

The draw at the presses should be straight, firm, and without flutter. A draw problem may exist due to one or more of the following faults:

- Soft stock
- Poor formation
- Wet sheet or wet streaks
- Dirty, worn, or partly filled press felt
- Filled press roll grooves
- Filled suction holes
- Improper press geometry or roll crown
- Improper roll or felt surface

A sign of trouble can be manifested in increased draw. Note the total or percent draw on the readout and how it compares to normal readings.

Vacuum on suction rolls and felt suction boxes are important. Higher than normal vacuum indicates a dirty or filled felt, or filled holes in a suction roll. Lower than normal vacuum indicates improper positioning of the box in the suction roll, air leaks at seals, or a malfunctioning vacuum pump. On a water ring vacuum pump, proper adjustment of seal water flow and temperature should be checked. Excessive roll cleaning shower flow or sheet moisture can cause a wetter press felt, which can raise the vacuum level.

The position of boxes in suction rolls is also critical, and this is important information to log. The leading seal strip must be in front of the nip to draw air and water out of the sheet and felt. The trailing strip must be beyond the nip or the sheet will crush because water expressed at the nip will be forced out of the nip against the fibers in the sheet. The suction box should also be positioned so that water stays in the suction holes after the nip. Water should stay in the holes long enough so it doesn't strike the felt when it is finally thrown out. This will help minimize rewet.

Top roll picking and excessive crumbs indicate a pitch condition or low sheet pH. A sheet that will not release is often caused by the top roll being too smooth. The top roll doctor condition should be examined often to ensure that crumbs and bits of broke are not wedging between blade tip and roll surface. Lubrication of the roll helps to reduce the tendency of crumbs to collect between the doctor blade and roll surface. A wet pad of broke remains along the line of doctor-roll contact. The roll can be lightly showered along this line of contact to remove this stock from the doctor.

Examine the in-going nip of the press for sheet blowing, which indicates a filled felt or filled holes in a suction roll. Blowing, of course, will result in press wrinkles and possible spot crushing. Press geometry may also be responsible for blowing. Puddling at the in-going nip indicates the suction roll box may be positioned to the dry end, the felt may be filled or too wet going into the nip, the holes of the suction roll may be filled preventing accumulated water expressed at the nip from being removed, or the felt or press roll may not be the correct design. Proper couching of the press rolls can sometimes correct this condition.

Press loading must match roll crown on conventional rolls. Nip loading should be uniform across the face of the rolls. Newer presses can have variable crown rolls that can be controlled. Crown on these rolls can be changed during operation. If actual loading is less than loading for which it was crowned, nip pressure will be light at the roll edges; if actual loading is greater, nip pressure will be heavy at edges and light in the center. Loading is occasionally adjusted, lighter or heavier, to correct for a cross-direction moisture or caliper variation at the reel. An error is corrected by creating a compensating error, two negatives make a positive. Avoid this practice because it affects sheet quality and results in uneven felt wear. Record monthly nip impressions, measuring nip width every foot. Graph the nip profile, and keep a history for every roll over its running life.

Checking of Primary Results

To maintain efficient press operation and uniform cross-machine sheet moisture at the reel, maintain the press nip loading and sheet moisture leaving each press uniformly. Sheet moisture leaving each press can be and should be controlled to ±0.5%. Sheet moisture meters (gamma gauges) have been developed and are a valuable tool to monitor sheet moisture on the run (see Chapter 8). This is a useful and important tool.

Another method requires breaking the sheet at the wet end of the machine. Collect samples of the sheet leaving the presses for moisture measurement. Take samples at multiple points, usually five, across the sheet width. This is a necessary check on the performance of each press. Take care to get samples that will provide a true measurement of moisture. The samples should not touch any part of the machine to avoid rewetting. Also, determine consistencies of water expressed at each press periodically. This check will determine press performance and degree of sheet crushing taking place. (Crushing will drive fines and fibers into the felt.)

Examine the sheet at the dry end of the machine for any indication of crushing, fiber realignment, washing of fines, and shadow marking.

The gamma gauge or microwave device can check felt moisture profile in the cross direction (CD) and machine direction (MD). This device can also determine the water removal at each press by the differences.

Monitor dryer steam pressure and steam flow. The back tender should advise the machine tender of any sudden or gradual change. Any change may be caused by some adverse condition of the machine or its operation other than at the press. Faulty operation of one of the presses, such as a worn or filled felt, filled holes of a suction roll, or improper press loading may be the cause.

Check presses mechanically for any vibration or roll bounce. This indicates roll misalignment, loose or out-of-round press roll cover, roll unbalance, holes of suction roll filled, or felt nonuniformity. Bearings and roll cover defects should also be checked.

Felt Performance

Examine press felts on-the-run at least twice each 8-h shift. If unable to correct problems, the machine tender should advise his supervisor.

For best press performance, felt tradelines should be straight. If the line is leading or lagging at one edge, that end of the stretch roll must be tightened or slackened. If the seam is leading or lagging at the center of the felt, one or two felt rolls having 90° or more wrap must be ground concave or convex to correct the problem. This can be done on a temporary basis by feeding a "leader" of kraft paper (wet-strength treated) around a felt roll to increase its diameter at the desired point. (See Chapter 6.)

There is a tendency to operate felts with too much tension. Excessive tension narrows the felt and reduces the voids in the weave. Instruments are now available for measuring and adjusting felt tension. After a run-in period, proper felt tension is maintained by running felts at their designed length. For methods of proper control, refer to Chapter 8.

A visual inspection of the felt indicates its condition and wear. If wear streaks are noticed at the edge, the fault is usually a poorly adjusted guide palm. A wear streak in the body of the felt or at an edge indicates a burr, a sharp edge, or an accumulation of pitch or other deposit on a slotted suction box or felt roll surface. A felt may have loose batting, yarn, or threads of yarn on its surface. These will result in press breaks, sheet marking, or sheet pickup of yarn fibers. Singeing the felt can correct this condition. Sometimes, a shanker wire is stretched across the width of the felt, close to the felt. Loose threads or felt fibers will strike this taut wire, which knocks them off the felt.

A tendency for the sheet to adhere to the felt at the exiting side of the nip is an indication that the felt-sheet combination is too wet, the surface of the felt is too smooth, or the press geometry is wrong. High surface tension forces are usually involved. The most common reason is a felt surface that is too wet. Felt compaction and filling, or a conditioning system in the felt run which is inadequate to dewater the felt properly, can create this problem.

Special problems occur on suction pickup felts. The sheet can have a tendency to fall away from the underside of a pickup felt. The cause of this may be a felt that is insufficiently wetted for the sheet to cling to the felt fibers by surface tension. A new bulky felt may not have enough felt fibers contacting the sheet to hold it. After a short run-in period the felt will compact and more felt fibers will be in contact with the sheet. A filled felt will prevent the vacuum at the pickup roll from lifting the sheet uniformly from the wire surface. Sometimes a sheet that is too dry will not adhere to the felt.

Occasionally, a wad will be lodged at the in-going nip during a press break. If the machine tender does not separate the rolls fast enough, the felt will be burned in a streak. A temporary measure may be to treat the felt with a bronze wire brush to clean and open the felt somewhat until the felt can be changed.

Seamed felts are available for presses. These allow faster felt changes and the use of heavier-weight felts. Seams should be checked twice a shift as a routine felt check.

Shoe-Type Press Operation

It is extremely important to prevent wad damage in the shoe-type press. The intense high pressure (6000 pli) can severely damage felts and rubber blankets, if wads are not kept out of the unit. Any broke between the blanket and the press felt can wear the belt, which causes a wet streak and eventual belt failure. Do not get water on the felt while loading the press. If a break occurs and it becomes necessary to wash the felt, be sure to take the load off of the press. If this is not done, the water creates such an intense pressure within the nip that the blanket can explode off the press.

The oil system is an important piece of the shoe-type press, so make sure the drain lines are kept free and open. Once a blanket is destroyed on the press, pieces of it get into the oil reservoir. It is important to keep pieces that could plug drain lines out of the oil reservoir. Clean and inspect the modules and strainers on a routine basis, and after every break.

A complicated series of interlocks on the hydraulic system of a shoe-type press requires expert troubleshooting knowledge. A list of interlocks available and their status is very helpful in troubleshooting. This list should be posted near the unit.

Performance of Press Auxiliaries

Felt roll and slotted suction box surfaces must be clean and free of any buildup, such as pitch, fibers, stock additives, aluminates, and felt fibers. Slotted boxes are now designed to permit cleaning on-the-run. Adjust felt cleaning shower flows for a minimum flow to prevent excessive wetting.

All shower water applied to felts, roll surfaces, and roll suction boxes should have an equal or slightly lower pH than the sheet. The temperatures should be controlled to prevent plating of pitch on the felts and surfaces contacting the felts and sheet. Check temperatures and pH of shower water and maintain within control points.

Examine felt travel for any water that may be accidentally wetting the felts.

Examine the felt guides to see that palms are properly centered and at right angles to felt edges. Check the position of the floating end of the guide rolls to be sure that they are in the midpoint of travel. This is accomplished by adjusting the opposite or fixed end of the guide rolls. The continuous felt cleaning system should be checked. Ensure that the felt cleaning system is functioning in accordance with its design and application. Proper shower flow, pressure, and angles are important. Proper operation of felt oscillation equipment is crucial. A stopped felt shower will wear streaks in the felt that can mark the sheet. (See Chapter 6.)

C. START-UP AND SHUTDOWN PROCEDURES

Examination and Cleaning

Whenever a press is stopped, even momentarily, relieve the loading. Run the press at crawl speed, examine the felt closely for wear, filling, or any damage, relieve the felt tension, and turn on the wash showers full. Even though continuous felt cleaning is employed, many mills clean the felts using specially prepared acid and alkali washes and detergents. If the press is to be down for any length of time, the press rolls should be separated. Raise the top doctor, and relieve the felt tension while the rolls are not operating. Examine the felt and press and clean the felt, as mentioned. This examination should include framing, roll bearing housings, saveall pans, doctors, platforms, walkways, and steps.

Accumulated dirt will eventually fall onto the felts, rolls, and sheet. Good housekeeping is an essential function of good machine operation. Press roll suction holes, grooves, and blind drilled holes should be checked. Doctor blades should be examined for wear and fit. Roll surfaces should be carefully examined for damage or wear. Showers, felt conditioning equipment, guides, and so forth should be cleaned and their operations checked. Gently tapping the roll with a rubber mallet can indicate a cover that is separating from the shell.

A machine tender should also use his ears to check press operation. Any strange noises or sounds can be important early warning signs of trouble.

Nip Impressions

Nip impressions are the primary indication of the press action on the sheet. Take nip impressions at each maintenance outage, or at least monthly. The impression should be done immediately after shutting the machine down while the rolls are still hot. Place a strip of plastic over the roll surface at the suction box on suction rolls. The nip impression should be taken with the vacuum on the suction roll. The nip impression can be taken with the felt on, if necessary. The roll should be loaded to normal running load with the nip impression paper in the full width of the nip. Observe how the rolls load to be sure they are coming down evenly. If one end appears to be lowering first, the full weight of the roll may be momentarily on this end. This can give a false reading on the nip impression.

Mark the nip impression paper every 12 in. across the width of the nip. At each mark, measure the nip width in millimeters from front to back. Record this information on a spreadsheet program where it can be saved and graphed. This becomes a permanent record over the life of the roll.

Impression paper is available that records calorimetrically the nip pressure, in psi. This can be helpful on an intermittent basis to identify a certain problem. It can be used dynamically while the roll is turning slowly.

Clothing Change

As mentioned before, felt runs and press arrangements are designed for simple, rapid clothing changes. Develop and write clothing change procedures. Train each person to carry out each specific action at the proper time. While the felt is off the press, take nip impressions of the press. Lay out the clothing to be installed in advance. Position the felt on the operating floor so that the installed felt will run in the proper direction. If the felt is two-sided, be careful that the correct side is contacting the sheet. It is good practice to locate the tradeline near press rolls when installing a felt to be sure the felt is installed straight. When putting the felt over and under press and felt rolls, it is good practice to place heavy wrapping paper over the machine parts to prevent felt damage by snagging it on a sharp edge or projection. The wrapper also prevents contamination of the felt with oil or grease. If oil or grease gets on the felt, remove it immediately with solvent.

After installing the felt on the press and everyone is clear of the press, turn the press at crawl speed. This stretches the felt to its running width. Lower the top press roll to a light nip contact, and position the stretch roll for a light felt tension. Tension should only be enough to drive all the felt rolls. Then inspect the felt closely for wrinkles, creases, or dirt spots. The next step is wetting the felt thoroughly using a detergent. During the wetting operation, full operating load is applied to the press.

Start-Up Procedures

After a press has been down for one hour to several hours, examine the press and felt before starting the press. Pay attention to the felt to be sure that it is clean and free of wrinkles. Lower the top roll onto the bottom roll and lightly load. Bring the doctor into contact with the top roll. Adjust the stretch so the felt has enough tension to drive all felt rolls. Don't load so much that the felt begins to rope. With the press at crawl speed, uniformly shower the felt with warm water over its entire surface. This is necessary to prevent uneven felt shrinkage and wrinkling. Shrinkage of synthetic felts is due to the tension on the felt in the MD.

When wetting is complete, load the top press roll to operating load and tighten the felt to normal operating tension. Turn off the wetting showers. Bring the press up to running speed. Run the press without a sheet for several minutes to allow the press to adjust to the running conditions.

When the tail is passed through the press section, adjust the draws to prevent the tail from pulling apart, or to prevent it from climbing the top roll and breaking when touching the doctor. Readjust the draws after the sheet is over the press.

Immediately after the sheet is over the machine, the machine tender optimizes the wet end of the machine - the headbox, former, and press part - for best operation, which includes safe running and production of a quality sheet.

Chapter 12

Practical Press Section Monitoring

A. INTRODUCTION

The use of instrumentation and other techniques for monitoring and trouble-shooting paper machine press sections is widely accepted. The need to produce at lower cost has emphasized the need for optimizing pressing efficiency. This chapter provides information on available instrumentation and procedures for monitoring wet press performance.

Trade names and manufacturers names are given, without value judgments, in order to address the practical day-to-day aspects of press section monitoring. General descriptions of the various measurements have been provided to assist the reader in selecting a supplier. These measurement principles are summarized in **Table 12.1**.

B. SAFETY

Paper machines, and particularly press sections, are complex machines that demand attention to personal safety. Chapter 15 discusses personal safety and should be reviewed prior to undertaking work or measurements in the press section.

C. FELT THICKNESS

Felt thickness indicates the degree of wear and felt compaction, which influences the specific void volume and permeability of the felt, and can be useful in predicting operational characteristics. Felt thickness continually changes during press section operation; the rate of change is strongly influenced by felt design. Generally, felt thickness decreases very rapidly in the first few days after installation, tapers off, remains nearly uniform for a period of time, and finally reaches a point where the felt is too thin to perform efficiently. Front side and back side measurements are taken while the felt is in operation and an average value is reported. Full cross-machine direction (CD) measurements may be made when the felt is off the machine, before or after use. A history of thickness information on each felt design is a useful indicator of when to remove a felt.

The caliper gauges for wet felt monitoring are lightweight, hand-held instruments capable of measuring thickness a few inches in from the edge of the felt. The basic instrument consist of a C-frame to which anvils are attached. A dial indicator reflects the separation of the anvils. The Huyck instrument uses a roller-type anvil and the Ames instrument a mushroom-shaped stationary anvil. The depth of the C-frame on the Huyck gauge is approximately 6 in. (15 cm). The Ames instrument can have a throat depth of up to 24 in. (61 cm).

Once the felt is off the machine, slits are usually made across the felt width to measure CD caliper after use.

D. SHEET OR FELT TOTAL MASS

Total mass measurements provide relatively accurate material balances around the felt loop to determine the impact of felt conditions on sheet dewatering. Cross direction total mass profiles of the wet felt can indicate uniformity across the machine width. Variations caused by improper press roll crowns, malfunctions in the press loading mechanisms, nonuniform showering and dewatering, and other sources frequently can be identified. Machine direction (MD) measurements are commonly used to detect roll bounce and MD felt nonuniformities.

Two types of total mass gauges for wet felt monitoring are available. The portable sensors offered by NDC Systems use a gamma radioactive source in the backscatter mode. The measuring probe is about the size of a three-cell flashlight and weighs about 2 lb (0.9 kg). Since total mass is measured by backscatter, gamma gauges cannot be used with rolls or other background objects within 12 in. (30 cm) of the sensor. As with any backscatter sensor, the total mass will vary with the distance of water or felt from the sensing probe. The gauge can be packaged in a case with a recorder.

The other basic type of total mass sensor, the transmission betagauge, is similar in principle to the unit used in monitoring the basis weight of paper at the dry end of the paper machine. The betagauges used for press felt monitoring incorporate a stronger source, however. The portable units (marketed only by Overhoff Technology) have the source and detector heads mounted on a C-frame. Betagauges had been the accepted technique for monitoring wet felt total mass, but their use has greatly diminished due to problems in gaining access with a C-frame to the felt runs in modern press sections. Also, their lack of portability compared to the gamma gauge has hampered their use.

Total mass measurements have been largely replaced by improved felt and sheet moisture gauges and are rarely used today. Government regulatory agencies control the transport and use of radioactive sensors.

E. FELT WATER CONTENT

Felt water content can be an important variable in some press systems, although numerous studies indicate no correlation between felt moisture and exiting sheet dryness in many commercial felt/roll systems. High water loads in the felt increase the hydraulic flow resistance of the felt in the press nip. This increases the inertial forces that have to be overcome to get water to flow out of the felt and the nip system and may increase sheet rewetting.

Usually, CD measurements are performed before and after each press nip (if possible) and before and after the felt suction boxes. The measurements are best made on the same side of the felt because some of the measuring devices preferentially detect surface water. Machine direction measurements usually are made prior to the uhle boxes, where the effect of nip bounces may be more easily detected.

As with total mass sensors, moisture measurements on press felts provide water balances around the felt loop and identify the source of CD profile variation. Most of the sensors available for water content measurements have fast response times so that short-term MD moisture variations can be monitored.

The most successful felt measurements have been obtained using instruments based on microwave and radio frequency (RF) principles. These sensors make use of the fact that water responds to these electrical frequencies differently from other materials normally present in the press section. Most of the instruments are small, very portable, and can be used at many different positions on the felt run, depending on press configuration.

The Scanpro JET MEM sensor has a built-in memory and does not require a cable connector to a central unit. This change has improved safety and portability. In addition to a standard recorder output, a RS-232 output is available for downloading information to a computer.

The M/K Moisture Sensor is a microwave sensor with a smaller measurement head than other sensors. The small size allows it to be placed in tighter areas of the press section.

Although all of these sensors respond to the water content in wet felts, absolute calibration on a wide range of felts can be very difficult. The microwave and RF sensors generally are sensitive to the distance between the sensing head and the water in the felt, which in turn varies with felt compaction and felt design. The measurements can be sensitive to operator use and to location on the machine for repeatable results. If possible, all measurements should be performed on the same side of the felt. The RF sensor is sensitive to water conductivity; a conductivity cell is supplied with the sensing head to permit measurement of and compensation for conductivity on the instrument calibration. Further development of these moisture sensors may make it possible to generate good calibration data on different felt designs.

F. SHEET WATER CONTENT

The moisture content of the sheet leaving a press is the basic indicator of wet press efficiency. A 1% increase in sheet dryness leaving the press section generally permits a 3 - 5% increase in production rate or a comparable decrease in energy consumption during drying. For pressing process studies, the ratio of sheet water to fiber is more indicative of pressing efficiency than sheet moisture content because sheet basis weight affects press dewatering.

To obtain the sheet water content, the web is measured before and after each nip in the press and the results are compared with the uncoated and unsized weight of the sheet. Most of these sensors require close proximity to the sheet while it is away from the felt, so care must be taken not to break the web. Unfortunately, the new, no-draw press designs have made this measurement increasingly difficult to obtain.

Techniques used to determine the moisture content of the sheet in the press section include grab moisture tests, total mass sensors, and radio frequency sensors. All except the grab moisture tests are performed with portable and on-line sensors.

The earliest and most widely used technique is the grab moisture test with subsequent weighing and drying in the laboratory. TAPPI TIS 0404-01, "The Determination of Water Removal by Wet Presses — Flat Paper Machine" outlines a test procedure.

The moisture content of wet webs can be determined by measuring the total mass of the wet sheet in the press section with a transmission betagauge or backscatter gamma gauge and subtracting the moisture-free mass of fiber, filler, and other additives. If surface sizing or coating materials are added to the sheet after the press section, their mass must be subtracted if the moisture-free weight at the reel is used as a basis weight for adjusting the wet press total mass reading. This procedure can lead to some inaccuracies, although the mass of water generally is substantially greater than the relatively minor errors which can be introduced by small changes in moisture-free weight at the reel.

One disadvantage of using total mass sensors on wet webs is the need for close contact with the wet sheet. The gap between the betagauge source and detector heads may be as low as 1 in. (2.5 cm), which can adversely affect sheet runnability. An advantage of betagauges is that their calibration in wet presses is generally easier than for other moisture sensors.

Portable infrared, microwave, and RF sensors used to measure sheet moisture content are generally the same as those used to measure felt moisture content.

ABB Automation, Measurex, and Valmet Automation are among suppliers that market on-line instrumentation to measure web moisture out of the press section. These sensors are generally infrared sensors mounted on O-frames, are not portable, and are offered as part of digital process control computer systems.

Most microwave, infrared, and RF sensors are reflectance units that do not transmit signals through the sheet. These units are generally sensitive to the distance between the web and the sensor head. Most of the sensors listed have been used as portable and on-line sensors with varying degrees of success.

G. DEGREE OF FELT FILLING/OPENNESS (POROSITY/PERMEABILITY)

The degree of felt openness is an important factor in the hydraulic flow resistance in the press nip. A filled felt restricts water migration in the nip and reduces water removal efficiency. Measurement of felt air flow properties during operation are useful in predicting performance, monitoring felt cleaning, and scheduling clothing changes. The measurements are made at intervals across the felt width to indicate areas of wear, filling, and other problems.

The instruments that monitor felt permeability are based on the flow of air or water through the felt. Air flow-based instruments include the Huyck-Smyth Porosity Tester (HSPT), the Frazier Air Permeability Tester, and the Appleton Mills Felt Air Flow Gauge. The Noram-PAPRICAN Felt Permeability Meter is based on water flow through the felt.

The HSPT is a portable suction box calibrated to yield an openness figure. The instrument is connected to a mill vacuum supply and held in contact with a running felt. The resistance to the air flow is measured in inches of mercury. The use of the Huyck-Smyth Porosity Tester is the basis of the TAPPI Suggested Method T 680 "Porosity Measurement of Press Felts on the Paper Machine (HSPT)".

The Frazier Air Permeability Tester is a standard laboratory instrument used extensively in the textile industry. This test measures the cubic feet per minute of air passing through a square foot of fabric under an applied pressure drop of 1/2 in. (1.3 cm) of water. The Frazier test has been adopted as a TAPPI Classical Test Method (T 251 "Air Permeability of Porous Papers, Fabrics, and Pulp Handsheets") for wet felts. Most felt manufacturers are equipped to measure and report Frazier test results on the dry felts they supply as well as to provide a testing service on returned felt samples.

The Appleton Mills Felt Air Flow Gauge is a turbine air flow meter that is placed over the top of the felt suction box. The volume of air flow passing through the felt and the vacuum level are usually reported. Because very few machines have identical vacuum capacity, this measurement cannot be used to compare machines.

The Noram-PAPRICAN Felt Permeability Meter is based on water flow rate through the felt. The instrument is placed over the felt and a jet of water is passed through the felt. The water flow rate is then reported, along with the water pressure at which the test was made.

Some mills use the uhle box vacuum as an indicator of the degree of felt openness. When a liquid ring pump is used to generate vacuum, the vacuum on the uhle box will increase as the felt becomes filled. Information can be misleading if several vacuum elements are supplied by the same vacuum pump.

H. RUNNING FELT LENGTH

Press felt running length must be within a specific size range to properly fit each press section. Woven felts were designed to run at a specific length for optimum performance relative to width, drainage, and tension. Modern felts are designed to operate in a specific tension range for best performance.

Running felt length is usually determined with a felt measuring wheel, which consists of a rotating element (usually 1 ft. [30.5 cm.] in circumference) connected mechanically with a footage meter. Buttons are incorporated to start, stop, and zero the counter. Some press sections have permanent position indicators mounted on the jack-roll take-up mechanism to allow for direct reading of running felt length.

A less-accurate method is to compare felt revolutions per minute (rpm) to press speed.

Many mills use the position of the stretch roll to chart the running length vs. time. This method can be used on felts running at a constant tension and provides a good indication of felt life.

I. FELT TENSION

Uniform felt tension is necessary for good drainage, optimum felt life, and consistently good runnability. Every felt is designed to operate in a specific tension range. Proper felt tension generally will be obtained if the felt is run to the desired length. Differences in felt stiffness may require different operating tension targets for varying felt designs.

Many machines now have pneumatic indicators or load cells which monitor tension and automatically control tension.

A Huyckmeter is a portable instrument capable of measuring paper machine felt operating tension. The Huyckmeter consists of a frame with two feet, a movable weighted arm and foot, a deflection gauge to measure the position of the movable foot, and a handle to permit movement of the instrument on a felt. The movable foot is positioned between stationary feet and it seeks a level depending on the felt tension. The deflection gauge is calibrated to convert the change in foot position to the change in felt tension. The application of the Huyckmeter is limited to press positions with an accessible, horizontal (+/-15°) felt run. Other gauges based on load cells are available from various instrument suppliers.

J. TRADELINE DISTORTION

Tradeline or weave distortion is a common phenomenon on paper machines. The tradelines are on the felt to show the position of the filling (MD) yarn in relation to the warp (CD) yarn. Any distortion of the tradeline means a similar distortion in the weave which can cause the felt to close up nonuniformly. A small amount of tradeline distortion is probably not significant, particularly if the distortion is uniform, but a tradeline that is off 1 ft (30.5 cm) or more for each 100 in. (2.5 m) of felt width changes permeability characteristics. Distortion of this magnitude may influence water removal, the rate of filling with foreign materials, felt and sheet moisture profile, and the felt stability.

K. FELT AND ROLL SPEEDS

Variation in drive speed on the press or between driven components on a single felt can be a source of pressing problems. Comparing felt surface speeds at different positions in the felt loop can be beneficial in trouble-shooting press sections. Portable electronic draw/speed indicators are used for accurate monitoring of roll and felt speeds. This type of unit consists of a transducer wheel which is held in contact with the moving surface and related electronics. The transducer generates a signal to the electronics relative to the surface speed of the roll or felt. When two transducer wheels are used at different positions in the press section, the differential between the measurement points can be read instantaneously.

Felt and roll surface speed measurement analysis should recognize that surface speeds can vary significantly across the machine due to speed differentials caused by crowned press rolls and concave or leadered felt rolls. Roll surface speed measurements taken outside the nip can differ significantly from the speed in the nip due to rubber deformation. The softer the cover, the larger the error; in a hard roll/soft roll combination differences can be large, depending on which roll is measured. Strobes and optical tachometers are also used to determine roll rotational frequency and speed. Many machines have permanently installed press roll speed measurement equipment.

L. PRESS VIBRATION

Trends to increased paper machine speeds, harder press roll covers, and higher nip loadings have led to a rise in the incidence of press vibration and roll bounce problems. Both terms refer to an abrupt displacement of the movable press roll that disturbs the normal level of nip loading. Press roll bounce is usually a low-frequency disturbance with a frequency of up to 200 cycles per minute (cpm). Press roll bounce generally can be observed and timed with a stopwatch. One or two nonuniform areas in a press felt, an irregularity in the cover of the press roll, roll misalignment, and unbalanced rolls, are common causes of press roll bounce.

Press vibration is usually measured in terms of cycles per second (Hz). The high frequencies require sophisticated vibration analysis equipment to characterize press vibration. Nonuniformities in felts, press rolls, bearings, and drives have been related to vibration.

An accelerometer or a displacement type pickup with a frequency range of 10,000 -20,000 cpm is used with most vibration analysis equipment. The vibration pickup feeds a signal into the vibration analyzer, which indicates both the frequency and amplitude of the vibration and allows the predominant frequency to be determined. The analyzer contains a tunable filter against all but the predominant frequency so that the pickup can be used to locate the source of the vibration. Strobe lights or optical tachometers triggered by reflecting off a roll may be used in conjunction with the vibration transducers in an attempt to correlate roll rpm with the predominant vibration frequency and thus narrow the vibration to a specific roll. A high-speed, multichannel tape recorder can record the signals from several points in the press section for use in later analysis. A high-speed recorder with a frequency response up to 200 Hz and chart speeds up to 8 in. (20 cm) per second is used to graphically display the measured frequency and amplitude. Some vibration measuring systems are permanently mounted to the press, allowing constant monitoring of press roll and felt instabilities.

Portable press roll analyzers yield information on press roll eccentricity and corrugations. These units make use of synchronized averaging of the signals from an infrared tachometer and a displacement transducer mounted on the same roll.

Some low-frequency press vibrations may be spotted using high-frequency MD felt moisture profiles. However, the more complicated vibration analysis equipment must be used in order to separate vibration from the other process effects in the analysis.

In addition to portable instrumentation, many machines have been equipped with on-line vibration measurement that is useful for scheduling felt and roll changes and maintenance.

M. PRESS ROLL ALIGNMENT

Misaligned press rolls can cause guiding problems, wrinkles, streaks, excessive wear on felts, and press vibration. Proper roll alignment provides uniform felt and sheet tensions and prevents felt roping and streak problems.

An alignment gauge offered by Appleton Mills provides a simple and effective method of determining the relative alignment of two rolls. The gauge consists of a telescoping rod, an adjustable level, and a micrometer. The gauge is placed against the face of two rolls approximately 6 in. (15 cm) from the front edge with the micrometer set at zero. The level is adjusted until the bubble is centered, then the gauge is moved to a position 6 in. (15 cm) from the back edge of the rolls. If the bubble remains centered, the two rolls are properly aligned. If the bubble is not centered, the micrometer is adjusted until the bubble is centered. The micrometer reading indicates the degree of roll misalignment. Repeat tests are conducted at 12 in. and 18 in. (30 cm and 45 cm) on both roll ends. Roll alignment can also be checked using engineering transits and accelerometers with frequency spectrum analyzers.

N. PRESS ROLL CROWN

The crown of press rolls refers to the shape or diameter profile necessary to compensate for deflection. This compensation is necessary to obtain a uniform nip pressure distribution across the width of the machine. Deflection is caused by sagging of the roll body due to its own weight and external loading factors. Roll crown definitions and specifications are discussed in TAPPI TIS 0420-06 "Roll Crown Specifications and Definitions".

Proper crowning is essential to producing a quality sheet while providing maximum roll and felt life. Improper roll crown can reduce operating efficiency due to nonuniform water removal across the entire face of the roll under a given nip condition. Roll crown also affects felt performance: Too much crown in a press roll will cause the tradeline in the felt to bow ahead in the center. This causes distortion resulting in plugging of the felt, uneven water removal, and possible stability problems. Too little crown will cause the tradeline to bow behind in the center, resulting in the same type of felt distortion.

A routine program of monitoring nip uniformity will maintain proper roll crowns. Static nips can be checked for width uniformity by placing nip impression paper in the nip, with the felt out, and closing the press nip. The impression is measured at several positions across the width of the paper machine and compared for uniformity. Special nip impression kits can be purchased from many felt and roll cover suppliers. Plain paper with a carbon backing may be also used. Press roll nip impressions should always be taken at roll operating temperature to ensure that the results are representative of normal operating conditions. Suction presses should be loaded higher than normal when the nip impression is taken to compensate for the additional loading that occurs under vacuum.

Fuji Prescale is an encapsulated film combination that measures actual nip pressure. The film is used in a similar manner to standard nip impression paper, except that the film develops a color density proportional to the pressure applied. The pressure is read by using a color density chart supplied with the film or by using the Fuji Prescale Densometer. This film has been used in rolling nips and can give information on roll cover corrugations. Static nip impressions are not always indicative of nip conditions if there are problems in the loading mechanisms.

Press felt and wet web CD profiles monitored with moisture and total mass sensors can also indicate press nip uniformity under operating conditions.

Variable crown rolls in the press section increase the flexibility of crown changes and lessen the dependence on grinding for a specific crown shape, but increase the monitoring requirements.

O. PRESS ROLL TEMPERATURE

Roll temperature is important to cover and roll life and nip characteristics. A covered roll that is run at increased temperature deforms more than one run at a lower temperature. If run too hot, a cover or roll can fail. Stream showers and high stock temperatures have made temperature monitoring important for efficient press performance.

Roll temperature measuring instruments can be divided into three areas: measurement of the water or oil cooling temperature in and out of the press roll; roll contacting measurement; and noncontacting measurement of the roll surface temperature.

Contact roll temperature measurement is accomplished with a thermocouple device placed against the roll surface. Friction is not a problem since the press roll is usually wetted and lubricates the instrument.

Noncontact infrared pyrometers are available as both single point measurement devices and thermal video systems. The single point measurement device, which shows the temperature when pointed at the roll, is easiest to use. Video systems are expensive and complicated, but can yield graphic information across an entire roll surface, as well as provide a video tape of the roll surface temperature over time. With infrared measurements, it is important to know cover emissivity for instrument calibration.

P. FELT AND SHEET TEMPERATURE

The sheet temperature is an important factor influencing the water removal in the press section. Measurements of the felt or sheet temperature profile across the machine have been useful in looking for moisture streaks. Wetter areas on the felt will be hotter if the sheet temperature is hotter than the rest of the press, and cooler if the sheet temperature is cooler than the press. The results must be interpreted carefully.

Noncontacting infrared pyrometers are usually used to monitor sheet and felt temperature.

Q. DEWATERING FLOW RATE

The amount of water removed from the felt conditioning devices and the press nips provides important information about how and where the water is being removed from the press. With this information, a water balance may be done on the press section. Vacuum felt conditioning dewatering flow rates can be measured by collecting the water from the press roll in a saveall pan that discharges through a weir.

Design considerations for common weirs and tables for calculating flow from the liquid height are contained in several standard engineering references.

R. FELT FIBER SHEDDING

Batt fiber that sheds from press felts during use can seriously affect dewatering capability and product quality. Techniques used to vacuum the sheet of finished paper on-line quantify the amount of batt fiber leaving the machine with the sheet. Either side of the web can be measured if proper precautions are taken. An apparatus produced by Huyck consisting of a narrow slotted measuring head (24 in. wide), a vacuum hose, and a removable fine wire mesh filter has been used to measure batt fiber shedding. This device connects to a vacuum source and the sheet is vacuumed for a set period of time. The debris caught by the filter is examined under magnification. Batt fiber is separated from wood and other debris, and a fiber count per unit time is produced. Particles can be tested with a drop of concentrated formic acid (use proper safety precautions) to determine their source. Nylon batt fibers will be immediately dissolved by the acid, while wood fiber will only be wetted and will not dissolve.

S. OTHER VARIABLES

Many other variables affect wet press water removal efficiency and must be considered in a press section evaluation. These variables include:

1. nip loading
2. machine speed
3. felt construction
4. roll cover construction and hardness
5. nip width
6. properties of the incoming web (freeness, water retention value, chemical additives, temperature, moisture content, basis weight)
7. type of press (plain, grooved, blind drilled, fabric)
8. suction box position and vacuum
9. suction box lubricating shower water pressure, and uniformity of coverage
10. suction or blind drilled roll hole blockage (groove plugging on grooved roll)
11. grooved roll purge shower pressure and nozzle condition
12. felt cleaning shower conditions (pressure, nozzle blockage, water pH, water temperature, application time, chemical addition rates)
13. felt suction box vacuum level or other variables on felt conditioning equipment
14. press configuration effects on sheet rewetting and sheet blowing.

Table 12.1 Wet press measurements

PARAMETER	MEASUREMENT UNITS	AVAILABLE INSTRUMENT TYPE	INSTRUMENT SUPPLIERS*
Felt thickness	Length	Micrometer	Ames, Huyck Felt
Felt total mass	Mass per unit area of felt	A. Radioactive backscatter gauges B. Transmission betagauges	A. NDC Systems, Overhoff Technology B. Overhoff Technology
Felt water content	Mass per unit area of felt	A. Microwave	A. M/K Systems, Scanpro
Sheet water to fiber ratio	Mass per unit area of sheet	A. Grab moisture tests B. Total mass sensors C. Microwave moisture sensors D. Infrared moisture sensors	B. ABB Automation Measurex, Overhoff Technology, Valmet Automation C. M/K Systems, Scanpro D. ABB Automation, Measurex, Valmet Automation
Degree of felt filling / openness	Percent available air, vacuum, or air flow	A. Porosity testers B. Suction box vacuum gauges	A. Frazier, Insco, Appleton Mills Albany B. Many suppliers
Running length of felt	Length	A. Portable indicators B. Permanently installed indicators C. Calculation from press speed and felt revolutions	A. Many felt suppliers and others
Felt tension	Force per unit felt width	Tension meter	Huyck Huyckmeter many instrument suppliers
Felt tradeline length and distortion	Length	Tape measure	

(Table 12.1 Wet press measurements, *continued from page 142*)

PARAMETER	MEASUREMENT UNITS	AVAILABLE INSTRUMENT TYPE	INSTRUMENT SUPPLIERS*
Felt and roll speed	Length per unit time	A. Electric draw/speed indicators B. Hand-held tachometers C. Optical tachometers	A. Avtron, Dynapar, and Others B. Many suppliers C. Many suppliers
Press vibration	Variations per unit time (length, velocity, or acceleration)	A. Alignment gauge B. Accelerometers with frequency spectrum analyzer	A. Appleton Mills B. Many suppliers
Roll alignment	Length	A. Alignment gauge B. Accelerometers with frequency analyzer	A. Appleton Mills B. Many suppliers
Press roll crown	Length Mass per unit area	A. Nip impression paper B. Radioactive backscatter and transmission C. Microwave moisture sensors D. Infrared moisture sensors	A. Fuji Prescale and others B. NDC Systems, Overhoff Technology gauges C. M/K Systems, Scanpro D. ABB Automation, Measurex, Valmet Automation
Roll temperature	Temperature	A. Roll cooling fluid temperature B. Roll contacting measurement C. Infrared single point D. Infrared video system	B. Many suppliers C. Many suppliers D. Flir Systems Kodak, Hughes, and others
Sheet and felt temperature	Temperature	A. Infrared single point B. Infrared video system	A. Many suppliers B. Flir Systems Kodak, Hughes, and others

**Addresses can be found in the Annual TAPPI Membership Directory and Company Guide. Any company wanting to be listed or deleted should send the appropriate information to Technical Association of the Pulp and Paper Industry, P. O. Box 105113, Atlanta, GA USA 30348-5113, TWX/FAX (404) 446-6947.*

Chapter 13

Press Section Economic Evaluation

A. INTRODUCTION

The economics of water removal has been a subject of interest in the paper industry for many years. In general, the cost of water removal increases in each successive step as the sheet advances from the head box to the reel. Drainage on the wire is quite inexpensive; wet pressing costs are more significant; and drying costs represent one of the major expenditures in the production of the sheet. The press section represents the last point at which water can be removed from the sheet mechanically, so it is important that uniform methods be developed for evaluating, comparing, and optimizing the performance of paper machine press sections in terms of economic factors.

B. PRIMARY PAPER-RELATED VARIABLES

Production Rate

In view of the high capital investment within the paper industry, production rate on a given paper machine is of prime concern. In any economic evaluation it is important to tie the results as closely as possible to the production rate. For example, press clothing life or even press clothing cost per ton of production is not nearly as important as the rate at which that production was achieved. In view of the fact that production rate often varies with grade being produced, it is desirable to determine an average daily production rate.

Sheet Dryness

Probably the single most important measurement for evaluating the economics of a press (or press section) is the dryness of the paper web as it leaves the press (or press section). In view of the higher cost of energy for drying paper and the effect of press section performance on drying costs, even a half percent in sheet dryness is economically significant. Not only is drying cost influenced by the moisture content of the sheet leaving the press section, but it can also affect the production rate for dryer-limited paper machines.

Whether obtained by appropriate instrumentation or by grab samples of the sheet, accurate web moisture contents are a must for any economic evaluation of the press section. A suitable means of collecting press moisture data is outlined in TAPPI TI Sheet No. 0404-01.

Downtime

Due to the capital-intensive nature of the paper industry, idle machine time is extremely costly. Production is lost (and therefore profit on that production) and the fixed costs involved in operating a production unit continue to accrue.

The cost of downtime for a given paper machine producing a certain grade of paper is often difficult to determine, and the engineer would be well advised to consult with the mill's cost accounting department in developing these data. The methods used to determine downtime costs vary widely from mill to mill, but the following factors are often considered:

- Depreciation on buildings and equipment
- Taxes and insurance costs
- Wages of paper machine crew (including labor overhead)
- Fixed portion of utilities costs (steam, electricity, gas, etc.)
- Lost profit (often expressed as standard profit per ton for a given paper machine and grade of paper)
- Office personnel, miscellaneous stores, general overhead

Regardless of the factors included in the analysis, for any given situation the downtime costs can be reduced to an hourly charge, D ($/h). This figure is useful in evaluating the cost of interrupted production for press fabric installation, sheet breaks, press fabric cleaning, and so forth, and is therefore rather basic to the evaluation of press or press section economics.

C. OPERATING COSTS

Press Clothing Costs

The cost of a press fabric, in terms of dollars per ton of paper produced, is a relatively easy figure to derive, but it can be quite misleading. Many factors and many interactions must be considered when determining the cost vs. benefits of a given fabric design. For example, a fabric with a long life but relatively poor water removal capacity is often a poor choice in terms of the total economics of wet pressing. Because of the economic leverage in wet pressing, the clothing cost must be tied closely to rates of production or water removal. This is particularly important for those paper machines which are production-limited by their drying capacity.

Press Fabric Costs

The specific cost of a press fabric is easily derived from its total cost, its life on the paper machine, and the average daily production rate during its life. This basic cost should be modified by the following considerations.

Fabric Installation Costs

The cost of installing a new press fabric should be included in the total press fabric costs. Some fabrics go on the machine quite easily while others are difficult to install, so this cost can logically be attributed to the fabric.

Fabric Cleaning Costs

Under consideration here are the cost of chemicals for continuous and batch cleaning as well as the cost of downtime for special washups attributable to the press fabric.

Maintenance Costs

The cost of maintenance is often difficult to allocate, but with proper maintenance records a reasonable estimate can be made. Depending on the detail required, it may be desirable to subdivide maintenance labor into categories according to the item being serviced (rolls, vacuum boxes, pumps, etc.).

Water Costs

In some mills the cost of the fresh and clarified white water used in sprays and showers in the press section can be a significant economic factor.

Cost of Lost Production

While there are innumerable possible causes for lost production or downtime on a paper machine, those which can be attributed to the press section should properly be allocated to pressing costs in any economic evaluation. Such problems as poor pickup, blowing, crushing, and so forth can be logically assigned to the press section. To fairly allocate these costs, a careful record of downtime on the paper machine in question would be required.

D. ENERGY COSTS

Press Drives, Wringer Drives, etc.

The electrical power supplied to the various press drives can be a significant economic factor in wet pressing. Most modern drives are direct current drives, so the power can be calculated easily.

For the case of line shaft drives or other types of mechanical drive systems, the problem becomes quite difficult and it is often necessary to resort to a series of estimates. For example, TAPPI TI Sheet 0406-04 (for cylinder machines) or 0406-05 (for fourdrinier machines) might provide a logical starting point.

Vacuum Pump Drives

The drives on the vacuum pumps can also be an important economic factor in the operation of a press section. The cost of vacuum for fabric conditioning, suction press rolls, pickup devices, and so forth can add up to a major energy consumption.

In the absence of direct power measurements, a rough estimate of vacuum power requirements can be obtained by using the rule-of-thumb that each horsepower of power input will generate an air flow of approximately 22 ft^3/min *(1)*. In addition, it is possible to estimate the power requirements from the total air flow and vacuum levels as suggested by DeCrosta *(2)*.

Water Pump Drives

In general, the energy inputs to these units will be insignificant compared to the above factors.

Thermal Input to the Sheet

The use of thermal energy from steam showers or other energy sources to increase the sheet temperature is an important means of improving press section operation. A heated sheet allows for more water removal in a press section because the viscosity of the water is reduced, allowing for higher flow rates of water from the sheet and press fabric. Elevated temperatures are associated with greater sheet and fabric compressibilities, thus enhancing water removal. Also, the benefits can be threefold: (1) drier sheet going to the dryer section; (2) lower heat input to the dryers because the sheet enters warmer; and (3) better moisture profile, or easier profile control. Because of the variety of techniques that may be employed, the economic evaluation of each installation must be tailored to local need.

E. POTENTIAL SAVINGS

Steam Savings

An increase in sheet dryness out of the press section results in direct savings of steam costs in the dryer section. The increased dryness may result, for example, from higher press loadings, furnish changes, or changes in the design of the press fabric. If all other pressing conditions are equal, the steam savings can be used to evaluate the economic benefits attributable to the single factor under consideration.

Savings Due to Increased Production — Dryer-Limited Machines

On many paper machines some of the heavier grades must be produced at lower speeds because of limitations in the machine's drying capacity. When this occurs and an increase in sheet dryness is produced by some change, it is possible to evaluate the economic benefits (including increased production rates) attributable to the single factor under consideration. To do so, a productivity factor, e_p, must be defined to account for the relief of dryer load brought about by the increase in dryness. While this factor will vary from machine to machine, as well as with the in-going sheet moisture, many mills find that a factor speed of 0.04 is about average. That is, a 4% change in machine speed is associated with a 1% change in sheet moisture, or sheet dryness. This relationship is approximate.

F. CLOSING REMARKS

The purpose of engaging in the manufacture of paper and board is twofold: (1) to manufacture a product, and (2) to sell it at a maximized profit with respect to the resources committed to the endeavor. Thus, the economics of pressing is a very important consideration, and it must be addressed with care and diligence. With the TAPPI resources, particularly the TI Sheets relating to papermaking, at the disposal of all, engineers and accountants should be able to approach the problem in a rational manner and work together to optimize the process.

References

1. Wicks, L. K. and Nelson, J. P., Pulp Paper 50(4): 78(1976).

2. DeCrosta, E. F., International Water Removal Symposium, London, 1975.

Chapter 14

Effect of Wet Pressing on Paper Properties

A. INTRODUCTION AND BRIEF HISTORY

The wet pressing section on a paper machine does more than remove water from the sheet. It can determine some of the physical properties of the finished paper. Specifically, these are caliper and density, strength, and surface properties. In the past, these connections were known to some extent but belonged more to the realm of the art of papermaking.

A great deal of R&D by machinery manufacturers and felt manufacturers has gone into removing more water. Press design considerations, roll covers, nip pressures, felt construction, and felt life have been changing continually to meet the demands of higher paper machine production speed and better paper quality. Surface properties can be greatly affected in the press, as can strength, stiffness, and fines and filler distribution.

B. FUNDAMENTAL MECHANISMS AFFECTING STRENGTH DEVELOPMENT

From a paper physics viewpoint, the apparent paper density is the single most important factor in determining the physical properties of a paper grade produced. Apparent density is measured by the basis weight per unit thickness. In the United States, apparent density is reported as lb basis weight/pt. of caliper (0.001 in.). Typical numbers are 3.0 lb (1000 ft^2)/pt. for a low density board; 11.1 lb (3300 ft^2)/pt. for writing and printing papers; and 4.2 lbs (1,000 ft^2)/pt. for high-density linerboard. The metric term is kg/m^3 and the conversion factor is lb basis weight/pt. x 192.4 = kg/m^3 for 1000 lb/ft^2 ream size.

Wet pressing directly affects the finished paper density by reducing the caliper at a point in the process where paper fibers are wet and flexible, and where fiber-to-fiber bonds are starting to form.

Wet density increase will improve paper physical properties and is not to be confused with caliper reduction (density increase) at the dry end of the paper machine with the calender stack. Dry end caliper reduction with a steel roll calender stack can reduce caliper, but this greatly reduces paper strength properties by breaking fibers and fiber bonds.

Bonding between fibers begins at approximately 22% solids. This means that bonding is taking place as water is removed in the press section and continues at an increasing rate as fibers are brought closer together. As the paper web is densified further, fibers are brought into intimate contact as the water is removed and bonding is increased by more intense wet pressing. The mullen strength, being a function of tensile and stretch, is also increased as solids content increases with wet pressing. Internal bond strength, as measured by the Scott internal bond test, increases with increased wet pressing. Another physical property used for measuring linerboard quality is the short span STFI Compression Index, which can be increased significantly due to a high sheet density as a result of very high wet pressing conditions.

C. SURFACE QUALITY OF PAPER

There is more to paper surface quality than smoothness as generated in the dry end machine calender.

It was discovered early in machine papermaking that the side of the paper last pressed against a smooth roll would produce a smooth surface on the finished product. For this reason, the reversing press was introduced. This led the sheet back through a nip where the wire side was pressed against a smooth roll in an attempt to smooth out the wire mark produced by the metal wire knuckles used at the time. (See **Fig. 4.3**, p. 31.)

Paper surface consolidation is an important effect of wet pressing on paper quality, especially for coated base papers. Good surface consolidation can be achieved by the action of the wet felt as it dewaters the sheet and knits the surface fibers together. This results in a strong paper surface and better coating pick strength tests. Some paper machines are designed with a fourth wet press specifically to consolidate the paper surface evenly, top and bottom side, to keep printing ink absorption even on both sides of the finished paper product. Pressing the paper against a very smooth roll just before entering the dryer section will result in improved paper smoothness on the side pressed against the roll.

The transverse direction in which water moves in a press is important for fines and filler distribution. The quality or nature of roll surfaces and press felt surfaces in contact with the paper will influence the surface details of the paper. Therefore, it is important that the press section design considerations be thoroughly thought out and engineered early in the preliminary design stage for a new paper machine or machine rebuild job. Paper end use and quality specifications will dictate the type of press section to be installed, and production requirements will determine how much water needs to be removed.

D. PRACTICAL ISSUES

A number of factors influence the effect of the press section on paper density. These include the number of wet presses, the maximum pressure applied (peak pressure in psi (i.e., pounds per lineal inch [pli]/ nip width = pounds per square inch [psi]), the type of press (roll press, long nip press, shoe press), the number of press felts (bottom felt only or double-felted nip), and the type of felt (batt on mesh, single- or multiple-layer weave, weight). Generally speaking, for most papermaking fibers the amount of sheet densification can be determined by the dryness (% solids) after the press section; that is, the drier the sheet, the more dense will be the paper at the reel.

It is important to note that excessive wet pressing pressure in the nip (hydraulic pressure due to filled felts, high water content in the felts, or other reasons) can reduce sheet physical properties by changing the formation. If carried to its extreme case, sheet crushing occurs which disrupts the entire fiber formation and results in holes in the sheet. Before crushing occurs, however, a period of incipient crushing may change the sheet structure and rearrange the fibers. Sometimes, paper surface and print quality changes may result.

Wet pressing can also impart shadow marking or shell marking to the finished sheet; these markings are caused by the press roll drilling pattern.

Chapter 15

Personal Safety Around Press Sections

A. INTRODUCTION

The information contained in this chapter is compiled from current safety standards.

The safety information given provides guidelines for press operators, students, designers and engineers and serves to remind them of the requirements for press section safety. This information supplements existing manufacturer's safety guidelines and existing safety procedures at the mill.

This chapter does not attempt to document every conceivable safety procedure or every unsafe practice.

B. PURPOSE

The purpose of the press section is to remove moisture from the paper or paperboard by squeezing the sheet between two rollers. To do this, machine builders use large heavy rolls and high nip pressures. The press section operates without interruption and meets the needs of different grades and properties of the paper and paperboard. It is essential that mills train all operating, maintenance, and supervisory personnel in press section safety and in the operating procedures for their particular press section. Personnel working on the press must never go into hazardous locations during press operations. Lack of proper training can be a cause of serious injury. The press section is between the forming section and the dryer section. Special safety measures must be taken at those sections with open draws. The operator should avoid transferring the tail by hand. Adequate barrier guards and safety signs must be installed to protect and remind personnel of the press section's potential hazards. (See paragraph 15/O for examples of unsafe practices.)

C. IN-RUNNING NIPS

PERSONNEL MUST NEVER COME INTO CONTACT WITH IN-RUNNING NIP POINTS. VERY SERIOUS INJURY CAN RESULT.

The following sketches illustrate various types of guarding to protect operators from in-running nips. Each guard must be sufficiently large and extend sufficiently beyond the nip to prevent access. An incorrectly placed guard could result in personnel entering the nip area from around the guard.

Because of the variety of press configurations and sizes, these sketches are examples only.

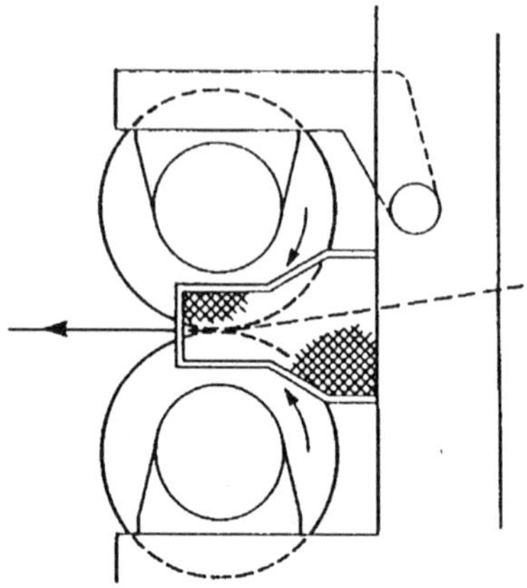

Fig. 15.1: Guard on a straight-through press

Figure 15.1 shows a simple guard on a single-nip straight-through press. The press frame provides additional protection to the in-going nip.

Figure 15.2 shows a simple guard on a single-nip straight-through press. Here the in-going nip is opposite from **Figure 15.1**. This requires careful placement of this guard because the nip does not have the added protection of the press frame.

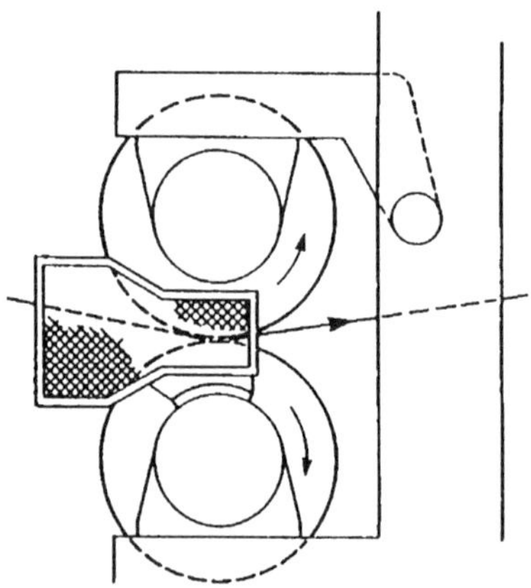

Fig. 15.2: Alternate guard on straight-through press

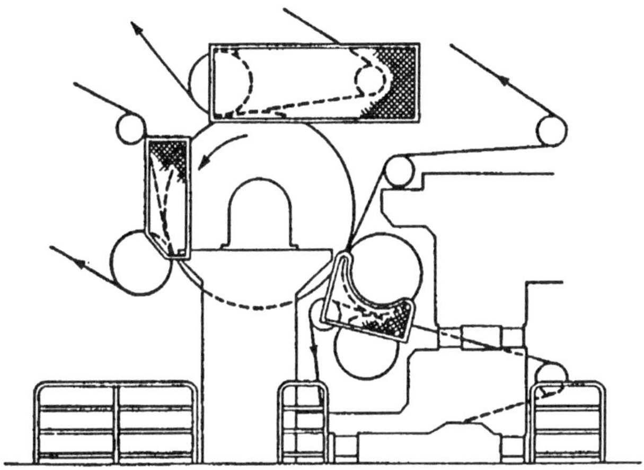

Fig. 15.3: Guard for direct action hot press

Figure 15.3 shows typical guarding for a direct action hot press. The in-going nips in many designs are similar to those shown in **Figs. 15.1 and 15.2**.

There are many press configurations ranging from the simple two-roll straight through press shown in **Figure 15.1** to very complex multiroll configurations such as the triple-nip and direct action hot press sections as shown in 15.3.

Manufacturers of press sections have the necessary expertise to place guards and recommend necessary safety procedures. All points of operation on the press within work and traffic areas must be protected by barrier guards.

A point of operation is the nip created by any two rolls. A work and traffic area is an area where personnel must be to operate the machine. Maintenance platforms are not work areas; they should not be used during normal operation of the machine and they should be chained off from normal access.

Consult the manufacturer for information regarding the placement of guards on your press section.

D. WRAPPING POINTS

PERSONNEL MUST NEVER COME INTO CONTACT WITH RUNNING PRESS FABRICS OR IN-RUNNING WRAP POINTS.

Besides the hazards created by the in-running nip of two rolls, there are other nips formed by a fabric travelling over rolls that must be considered. All wrapping points within the work area must be protected by guards or other means. Examples of guards are shown in **Figs. 15.4 and 15.5**.

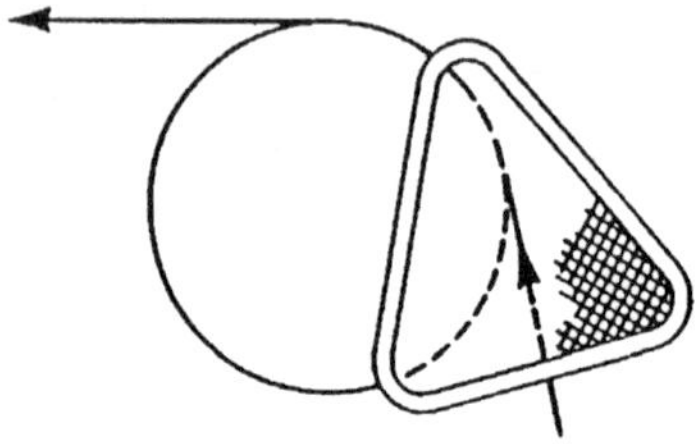

Fig. 15.4: Nip between felt and roll

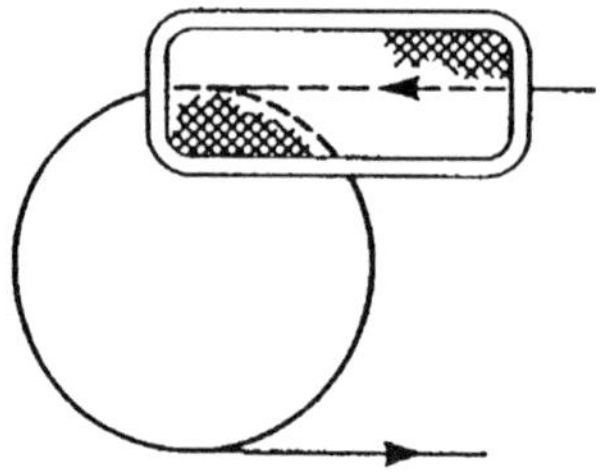

Fig. 15.5: Nip between felt and roll, alternate

Wrapping guards must protect personnel, their body extremities, and parts of their clothing from access to the nip. Sometimes, the placement of a handrail and footwalk at a safe distance from the nip will act as a separating device and prevent access.

Operators and maintenance and supervisory personnel must never touch a moving felt or roll under any circumstances.

E. FABRIC CHANGES AND ROLL CHANGES

Mill personnel must read and understand the manufacturer's operation manual for the proper safety procedures during fabric and roll changes. Always lockout-tagout (see paragraph 15/I) before performing any work.

Fabric and roll changes must be carried out with extreme caution. Controls should never be started until all personnel working on the press are accounted for and their location known. Be careful during fabric and roll changes to ensure that all slings, hooks, and other lifting devices are in good condition for handling fabrics and rolls.

Personnel must always use footwalks, access steps, platforms, and ladders to operate the machine safely, change fabrics and rolls, and do routine maintenance. Personnel must never stand on rolls or cross shafts.

F. FOOTWALKS, ACCESS STEPS, AND LADDERS

Personnel must always use footwalks, access steps and ladders as directed in the operator's manual. Do not climb onto machine framing or stand on rolls.

Where there is danger of falling into the machine from a footwalk, railings must be present on the machine side. Where there are footwalks or work places with a falling height of more than 40 in., a hand rail must be present on the open side.

During fabric and roll changes, portions of footwalks may be removed, thus disrupting the normal access routes. A guard rail must be provided to prevent personnel from falling off; chains generally do not prevent falls. Openings in railings for access footwalks must be guarded to prevent personnel from falling off. Such openings can be guarded by a door which only opens toward the work platform.

Ensure footwalks, hand rails, and guards are securely in place before starting the machine. All removable sections should be secured.

G. THREADING

Because of the many grades of paper and paperboard being manufactured and the large number of different press configurations, the following instructions for threading of press sections are given as general rules.

Threading by hand should be avoided. Always use threading aids such as air hoses, paddles, and rope threading systems. Always work on the outgoing nip side of the press.

All operating and supervisory personnel must learn and understand the threading procedure for their press section. Consult with the manufacturer of your press to assist you in establishing and developing current procedures for threading.

H. ROTATING SHAFTS

All transmission shafts such as main indrive and auxiliary drives within the section must be guarded according to Occupational Safety and Health Administration (OSHA) regulation Subpart R 1910.261 (b) (1).

I. LOCKOUT-TAGOUT

All personnel must be trained in the proper procedure for lockout (refer to OSHA Subpart J. 1910.147). Lockout and tagout devices shall identify the employee applying the device.

The main drive must be de-energized and locked out before any roll change, fabric change, or maintenance procedure.

All press rolls must be locked in up position and controls locked out before personnel work on the press.

After work completion, replace all guards, ensure no unsafe condition exists, and ensure that all personnel are accounted for and clear of the press before unlocking the lockouts and energizing the controls.

J. EMERGENCY STOPS

Press sections must be equipped with devices that will stop the press in an emergency. OSHA recommends that the devices consist of pushbuttons for electric motor power (or electrically operated engine stops), pull cords connected directly to the prime mover, control clutches, or other devices interlocked with adequate braking action. These emergency stops must be located so that in an emergency any person working on the machine can stop the machine section, or the entire machine, in the quickest possible manner.

Emergency stop devices and stop buttons or electrical switches with letters or other markings used for emergency stopping of machinery shall be red in color.

Quick stop devices should be tested periodically by using them during normal operations, or during planned shutdown periods to make certain they are always operational.

Emergency stop devices should not be relied upon as operating devices and should only be used for stopping the machine in an emergency.

All employees must be made aware of the emergency stops in their section as part of their safety training. (See OSHA 1910.261, Section K-1 Pulp, Paper and Paperboard Mills.)

K. OVERHEAD CRANES

Overhead cranes are used in the removal of rolls and during press fabric changes. Crane operators and other personnel must be aware of all safety procedures for the safe operation of overhead cranes.

Personnel working on the top press frames must be alerted to the hazard of the overhead crane travel. Crane operators must be aware of the possibility of workers in the way of crane travel and give warning to enable workers to keep clear.

Slings, hooks, and lifting devices must be checked periodically to ensure their good condition. Personnel must understand the proper procedures for the handling of felt rolls and felt clothing. Use the proper lift points to pick up and transport equipment as recommended by the manufacturer.

L. BROKE PITS

Pulpers underneath the press section must to so constructed that the broke quantities to be fed can be safely received. This requirement is met when the size of the pulper is designed according to the amount and grade of broke anticipated.

M. WARNING

Pulpers must be guarded with handrails. Gates must only open outwards and must only be capable of opening when pulper is locked-out. Do not place temporary working surfaces over an operating pulper.

If for any reason pulps have to be worked on, or if workers are repairing the press directly above the pulper, the pulper must be locked-out according to OSHA 1910.261 (j) (5) as follows:

(5) Pulpers. (i) All pulpers having the top or any other opening of vessel less than 42 inches from the floor or work platform shall have such openings guarded by railed or other enclosures. For manual charging, openings shall be sufficient only to permit the entry of stock and shall be provided with at least two permanently secured crossrails, according to American National Standard A12.1-1967.

(ii) When cleaning, inspecting, or other work requires that persons enter the pulpers, they shall be equipped with safety belt and lifeline, and one person shall be stationed outside at a position to observe and assist in case of emergency.

(iii) When cleaning, inspecting, or other work requires that persons enter pulpers, all steam, water, or other control devices shall be locked or tagged out. Blank flanging and tagging of pipe lines is acceptable in place of closed and locked or tagged valves. Blank flanging of steam and water lines shall be acceptable in place of valve locks.

(iv) All power mechanisms shall be guarded in accordance with American National Standard B15.1-1953 (reaffirmed 1958).

N. STEAM HEATED PRESS ROLLS

The warm-up procedure for steam heated press rolls must be strictly adhered to. This is necessary in order to avoid excessive stress concentrations in the press rolls due to large variations in temperature. Failure to follow proper procedure can be dangerous to life and machinery.

O. UNSAFE PRACTICES

To avoid injuries, operators and other personnel should be aware of and avoid the following unsafe areas or practices. Where unsafe practices exist consult the manufacturer of the press for proper procedure to follow.

- Unguarded nip points
- Unguarded wrap points
- Unguarded rotating machinery
- Unguarded moving members
- Poor maintenance of hoist equipment
- Improper handling of broke
- Inadequate barriers
- Unguarded broke pit
- Failure to lockout and de-energize when working on or repairing the press or press pulper and during roll and fabric changes
- Improper threading of machine
- Hand threading
- Inadequate safety signs
- Removal of handrails and guarding
- Improper use of footwalks, crosswalks, access steps, and ladders
- Poor housekeeping, failure to keep working and traffic areas free of broke and tripping hazards
- Improper storage of wash-up hoses
- Improper protection of doctor blades
- Improper care when cleaning with chemicals
- Improper warm-up procedure of steam heated press rolls
- Improper maintenance of hydraulic hoses and fittings

P. SAFETY SIGNS

Operators of press sections besides being protected from the hazards by a physical barrier must, in addition be warned of the hazards by the placement of safety signs. These signs must alert persons to specific hazards, the degree or level of the hazard, the consequences of the involvement with the hazard, and how the hazard can be avoided.

Consult the manufacturer of your press for the applicable safety signs and warnings.

In order for the hazard level to be easily identified, color coding for the signal words "Danger," "Warning," and "Caution" must be used. For example:

Danger– White letters with red background indicates an immediate hazard that if not avoided WILL result in death or serious injury. This signal word should be limited to the most extreme situation.

Warning– Black letters with orange background indicates a potential hazard that if not avoided COULD result in death or serious injury.

Caution – Black letters with yellow background indicates a potential hazard that if not avoided MAY result in minor or moderate injury. This sign is also used to alert against unsafe practices.

Safety Instructions – The safety instruction message is often made up of a number of individual messages stating procedures or actions that must be followed for the safe operation of a product. However, safety instructions do not take the place of DANGER, WARNING, or CAUTION labels.

The above definitions were taken from the ANSI Standard Z535.4 on Safety Signs and Labels. This Standard must be referred to when designing safety standards and labels.

Q. MAINTENANCE

Mill personnel must refer to the operation manual for proper safety procedures during maintenance. Because of the wide diversity of press design, it is impossible to cover all aspects of maintenance here. The following do's and don'ts for maintenance will serve as reminders.

- Do not walk on press or press fabric rolls. Use only the walkways provided.
- Block, pin, or lock-up to hold any equipment in its raised position as a precautionary measure against accidental energizing of controls.
- Follow the manufacturer's recommendations when removing bearings hydraulically.
- All services to the machine must be locked-out with a padlock before any maintenance, inspection, cleaning, or servicing is performed.
- The motor's main power disconnect switch must be locked-out.
- Check disconnect, try to start motor before proceeding further.
- All sources of power and material flow must be shut off, including bleeding off of pressure and locking out all pneumatics, hydraulics, chemical and gas systems, steam systems, electrical circuits.
- Never start the machine unless all personnel are clear of machine, all doors are closed, and all guards and cover are in place
- Aisles must be clean and clear of obstructions. Wipe up spilled oil, grease, unnecessary water, and stock. Good housekeeping prevents injuries.
- Do not stand on cross shafts.
- Do not work above an operating pulper.
- Do not stand astraddle of the machine line shaft or motor couplings.

R. OSHA REFERENCE

The following OSHA Regulations are listed below for your reference.

OSHA Code of Regulation 29, Parts 1900 to 1910 includes Part 1910 Subpart R - Special Industries.

Part 1910, 261 Pulp, Paper and Paperboard Mills

(a)		General Requirements
(b)		Safe Practices
(b)	(3)	Floors and Platforms
(b)	(4)	Lockouts
(c)	(8)	Cranes
(j)	(5)	Pulpers
(k)		Machine Room
(k)	(1)	Emergency Stops
(k)	(2)	Drives
(k)	(4)	Walkways

All machine operators, maintenance and supervisory personnel should not only read and understand the selected OSHA sections listed but all applicable OSHA codes pertaining to their job duties and functions.